STEP-BY-STEP

TELEPHONE INSTALLATION AND REPAIR

STEP-BY-STEP

TELEPHONE INSTALLATION AND REPAIR

JOE G. PENA

FIRST EDITION

FIRST PRINTING

Printed in the United States of America

Library of Congress Cataloging in Publication Data

Pena, Joe G.
Step-by-step telephone installation and repair.

Includes index.
1. Telephone—Amateurs' manuals. I. Title.
TK9951.P46 1986 621.386 85-22239
ISBN 0-8306-0984-9
ISBN 0-8306-1984-4 (pbk.)

Contents

Introduction

YOU CAN INSTALL YOUR OWN TELEPHONE SYSTEM! IT IS now legal. This book gives you the information you need to accomplish it in concise and straightforward steps.

For many years the "phone company" has taken care of all aspects of the phone system, but its responsibilities have become different since Bell System's divestiture from AT&T (January 1, 1984) and the resultant changes in FCC Rules and Regulations. The telephone company's only responsibility now is to bring a dial tone to the premise. This may seem like a terrible nuisance to you, but it opens the door to new opportunities and advantages. If you wish, the phone company can still handle the inside wiring, although you must pay extra. Your other options are to hire an independent company, or to do the work yourself.

Installing the phone system yourself has several advantages. It will save you money! You will have available the entire array of new equipment to please your personal taste! You can work at your own convenience, instead of waiting around for the service person! Finally, you will have the personal satisfaction of having installed it yourself!

The average person is quite capable of installing his/her own telephone system. The most difficult hurdle to overcome is fear: fear of not having the knowledge and fear of not having the skill. This book will lead you through each step of the way. You will not need to wade through pages of material that do not pertain to your situation. You can select the areas that are unique to your home. If you can, get a friend to help. It is always easier to decide how to proceed if you can talk it over with someone. Also, an extra pair of hands, though not absolutely necessary, can certainly be useful.

You should scan this outline before you read the book.

Read *Safety* (Chapter 1).

Read *Rules and Regulations* (Chapter 2).

Select and read the chapter that corresponds to your home.

Apartments (Chapter 3).

Homes Under Construction (Chapter 4).

Homes with Existing Service (Chapter 5).

Mobile Homes (Chapter 6).

The chapter you have selected will help you with the following:

Selecting method and routine of installation.

List of tools needed. Refer to *Tools* (Chapter 9) for details.

List of equipment needed. Refer to *Equipment* (Chapter 10), *Wire* (Chapter 7), and *Jacks* (Chapter 8).

Purchase, rent, or assemble tools and equipment.

Follow step-by-step installation for wire. See *Wire* (Chapter 7) if questions arise).

See *Jacks* (Chapter 8) for termination.

Troubleshooting (Chapter 11) will be useful to you even if you do not plan to install any telephone equipment. It will help you to identify, isolate, and solve your specific problems without having to read the entire section. If you have trouble understanding any terms in the text, the Glossary will give you a short clear explanation. I hope this book will answer the questions that you have about your telephone service and give you a virtually troublefree system. Best of luck.

Chapter 1

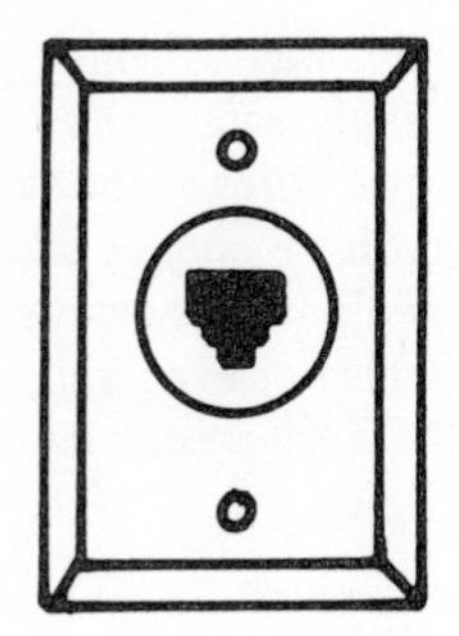

Safety

THERE IS NO REASON THAT ANYONE SHOULD BE HURT INstalling their own telephone. It is not a particularly dangerous thing to do. It is no more dangerous than cooking, or mowing the lawn, and certainly not nearly as dangerous as driving a car. Almost everything we do is dangerous if we are very careless or do not understand how to handle something. This section will help you to work carefully and safely.

Handtools are often used carelessly. You can hurt yourself badly performing a simple task with an ordinary handtool. Pay attention to what you are doing. Use the correct tool for the job. The wrong size screwdriver, for example, can not only make the job more difficult, it is more likely to slip and cut into your hand or other parts of your body. Check to see that your tool is in good working condition. Never force a tool! Never work at eye level. Always work moving the tool away from, rather than toward, your body.

Electrical handtools require a different kind of care. You may use an *electric drill* when installing your telephone. The drill should be the heavy duty type and should be in good working condition. Long hair should be tied back. Any loose clothing or jewelry that might be caught, should be removed. Do not operate electrical tools next to flammable liquids or in an explosive atmosphere. Use a drill that is either *double insulated* (printed on the side) or *three prong* (plugs into a grounded, three hole, outlet). If you do not have a grounded outlet, you will need to use an adapter. The green wire or the grounding lug will need to be connected to a permanent ground on the outlet.

The voltage for telephones is usually 48 Vdc and the ringing current is 105 Vac (low amperage). Phone currents are not normally dangerous. To be absolutely safe when working with telephone wires: (1) do not handle two wires at the same time (2) unplug the network interface (if you have one) *or* take one phone off of the hook. Phone company equipment, however is often located close to ac electricity that can hurt or even kill you. If there is any danger from electricity in the work area, *turn off* the power and check for hazards.

HINTS ON LADDER USE

In order to reach certain places you may need to use a ladder. These few simple precautions may prevent a broken bone or other injury. Never set up a ladder that is not on solid, level ground. If the ground is not solid, the ladder might start out level and then sink on one side under your weight. If you are using an a-frame ladder do not stand on the top or on the next step below the top because it is not stable. If you are using an extension ladder against the wall of a building, be sure that the distance from the wall is one-quarter the length of the extended ladder. The *fireman's position*, shown in Fig. 1-1, is a simple way of safely positioning the ladder. Stand straight, parallel to the wall, toes of your feet touching the feet of the ladder and with the ladder rungs at arms length from your body. This will work for any length ladder. When working on the ladder, never climb so high that the top of the rail is below your shoulders. Never lean out past the point where your breastbone is against the siderail or your body extends a foot past the siderail.

TO BE SAFE

1. Keep work area uncluttered and clean.
2. Do not work when tired.
3. Dress for the job. Do not wear jewelry and tie back long hair.
4. Keep children, pets, and visitors out of work area.
5. Use electrical tools only in a dry area.
6. Never use electrical tools next to flammable liquids or in an explosive atmosphere.
7. Use safety glasses.

Fig. 1-1. Fireman's position for the proper ladder angle against a wall.

8. Use correct tools for the job.
9. Keep tools in good condition.
10. Always ground electrical tools (or use double insulated).
11. Do not force a tool.
12. Do not overreach (keep both feet on solid surface).
13. Do not work at eye level or pull tools toward your body.
14. Put tools away when finished.

15. Persons wearing a *pacemaker* should not install their own telephone or engage in any other activity that could expose them to even a mild electrical shock.

Chapter 2

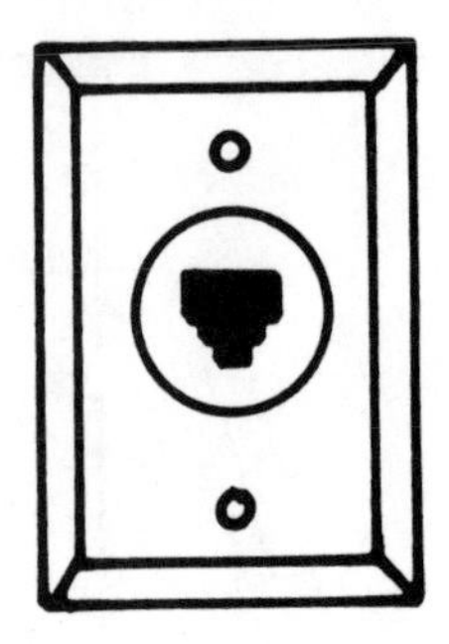

Rules and Regulations

IT IS NOW LEGAL FOR YOU TO INSTALL YOUR OWN TELEPHONE system. The telephone company is responsible for the outside line up to and including the interface or protector (point of demarcation). You are responsible for your inside system up to the point of demarcation.

There are certain guidelines that assure compatibility between your system and the larger telephone network. The Federal Communications Commission has established a system of registering equipment. Contact your local telephone company. They may have some rules of their own. They may need to be informed if you plan to open the protector and they will probably require you to register the FCC registration number and the REN (Ringer Equivalence Number) with them. Telephone companies use a system of codes for services and equipment called USOC. Some knowledge of this code may be useful to you.

FEDERAL COMMUNICATIONS COMMISSION (FCC)

Before you can install any telephone or telephone equipment, the FCC requires that you notify the telephone company. The telephone or equipment must be FCC registered in order to connect with their lines. This is easy to determine since approved equipment will have a *FCC registration number* and a *ringer equivalence number* on the equipment. The FCC registration number will have fourteen digits and letters. The ringer equivalence number, depending on the telephone or equipment, will read in whole or decimal numbers. The ringing current supplied for one telephone line (number) is 5.0. If the sum of the REN numbers of all of the equipment on one line (phone number) is greater than 5.0, the telephones will not ring.

Example:	phone 1	1.0 REN
	phone 2	0.8 REN
	phone 3	2.0 REN
		+
		3.8 REN

In this case you can still operate other equipment as long as its REN is not greater than 1.2.

USOC

No, USOC does not mean United States Olympic Committee in "Telephone talk." It is a term for the codes of services and equipment in the telephone system. The letters stand for *Uniform Service Order Code.* I have also seen it called *Universal Service Order Code.* These codes identify types of equipment used to provide telephone service. Below are some examples of jacks and adapters you are most likely to see in a residence.

JACKS

RJ11C—is the USOC for a standard desk modular jack. It can be a flush (outlet) or nonflush (surface) type. You can plug in any form of telephone equipment that only needs a single telephone line. This is the most common type for a residence.

RJ11W—is the same as the RJ11C except it is a wall mount type instead of a desk type.

RJ14C—is like a RJ11C except it provides two telephone lines. This jack can be used (with a special phone or a regular phone with an adapter) for a business or a residence to receive two different numbers on a single telephone.

RJ14W—is the same as a RJ14C except it is for a wall phone jack.

RJ25C—is used mostly on business systems that require three lines working from one jack. These come with six contact rails instead of the four in the standard modular jack. There is no difference in the size of the plug except for the pair of extra contact rails.

RJ31X—connects special telephone equipment ahead of all other telephone equipment. It cuts off all telephone equipment in front of it. It is most commonly used for automatic dialers in burglar alarms or answering machines.

ADAPTERS

RJA1X—converts a four-prong or four-pin jack into a modular jack. It makes a four-prong jack into a RJ11C. This adapter allows any standard modular cord to plug to the four prong.

RJA2X—converts a standard modular jack or RJ11C to an adapter into which you can plug two separate modular cords. This allows you to have two phones, or a phone and an answering machine, in the same place.

LOCAL

In addition to registering the REN with your local telephone company, you should also inform them if you plan to open the protector, and ask them about their own individual rules and regulations.

POINTS OF DEMARCATION

When you work on the telephone wire in your home, you must be able to separate the phone company wiring from your own residence wiring. Early equipment uses a protector as the demarcation point between the two. Later equipment still has a protector but it also has a network interface and the interface is the point of demarcation. If you have an interface, you should never need to open the protector. If you only have a protector you should obtain permission from your local telephone company before you work on it.

STATION PROTECTOR

The primary purpose of a protector is to defend telephone equipment from lightening or sudden electrical surges over the line. Most protectors defend one line (telephone number). A unit has two carbons to break the circuit and a ground wire to lead the voltage to the ground. This ground wire should never be removed or loosened. If for any reason it is, check the tag on the grounded end for directions. Some protectors defend two lines (two different telephone numbers) and a few defend even more. If your protector is for two lines (instead of one), it will have four carbons (instead of two) and four termination screws for two pair of wires (instead of two screws for one pair). The third or fifth termination screw is the ground.

Protectors come in all shapes. Usually they are easy to find because they are located beside the electric meter. Figure 2-1 shows one on the bottom right of an electric meter. If you do not find yours there, locate the drop (phone cable to your home) and trace it to the house. There are two types of drops: (1) aerial—provides a single telephone line and works from a telephone pole; (2) buried—can service two lines at once and runs underground.

Fig. 2-1. Protector mounted on side of home next to electric meter.

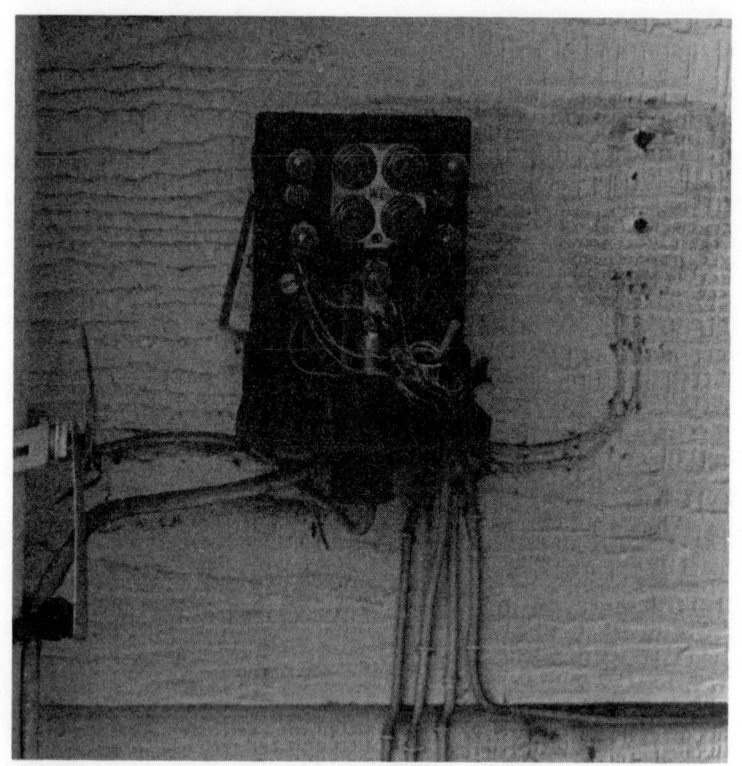

Fig. 2-2. Protector showing terminals for inside wiring and ground wire.

If you do not have an interface and must use your protector as the point of demarcation, you should call the local telephone company before you open it.

You can see in Fig. 2-2 that inside the protector, there are two screw posts on which to terminate the drop for your telephone line (number). The telephone company line is wrapped at the base of each screw. The inside wires (up to four pairs) are wrapped between the washers on the screws. (Each pair is split with one wire on one screw and the other wire on the other screw.) A nut is screwed on top to secure the wires and washers. Information on how to check the wiring on protectors is located in Chapter 11.

INTERFACE

Although interfaces come in a variety of shapes, there are really two types of interfaces: outdoor and indoor.

Fig. 2-3. An example of an outdoor network interface.

Outdoor Interface

Outdoor interfaces are usually located beside electric meters as in Fig. 2-3. If yours is not, locate the drop (phone cable to your home) and trace it to the house. There are two types of drops: (1) aerial—provides a single telephone line and works from a telephone pole (2) buried—can service two lines at once and runs underground.

Fig. 2-4. Outdoor network interface with cover open showing inner workings.

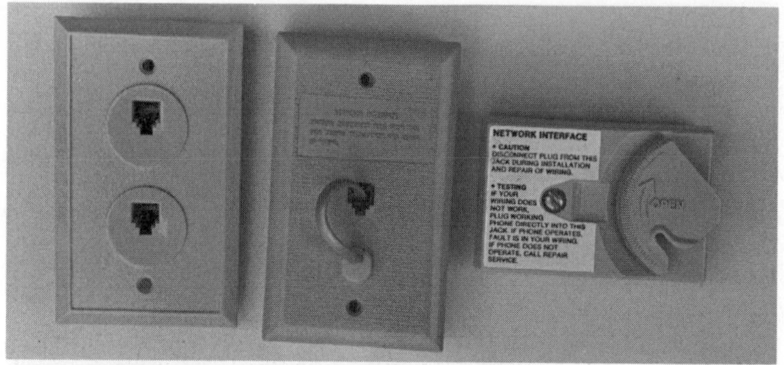

Fig. 2-5. Three types of indoor network interfaces.

An outdoor interface functions as the protector, the interface, and the wire junction for your inside wiring (i.e. you only need one box.) Figure 2-4 is an outdoor interface with the cover open to show the inner details. The customer wiring bridge, to which the station wires from the house are terminated with the corresponding colors, is located under the Customer Access Lid.

Indoor Interface

Indoor interfaces do not contain protectors (the protector is still outside). Some indoor interfaces have built-in jacks. Others need to be used in conjunction with a jack (with a short modular cord) or a wire junction (with a short modular cord). Figure 2-5 shows three types of indoor interfaces. The middle interface is the one with the built in jack. It contains the interface, the jack, and the modular cord. Nothing else is needed. The interface on the right is just an interface. It still needs a jack or wire junction, and a cord. It is in a moisture protected housing. The duplex jack, on the left of the figure, contains the interface and the jack but needs a short

Fig. 2-6. Surface inside interface with instructions on cover.

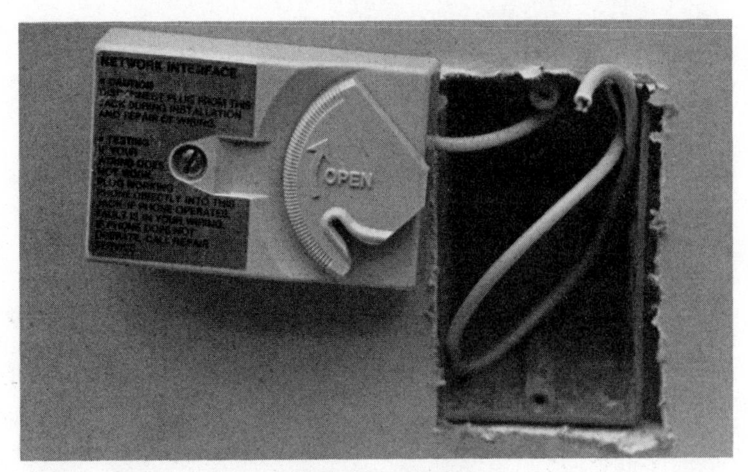

Fig. 2-7. Network interface installed in outlet box.

modular cord. Its original purpose was to supply two separate phones from one outlet but it can be used as an interface. The top part of the jack is wired to the outside, the bottom to the inside. A short modular cord will be plugged in, one end in the top outlet and the other end in the bottom outlet. Interfaces usually have specific instructions for their use. See Fig. 2-6.

Sometimes a prewired new home or apartment may have an installed interface that requires a wire junction, but the junction has not been installed. This will look something like Fig. 2-7. Install the junction as shown in Fig. 2-8. See Chapter 8 for wiring details.

Fig. 2-8. Network interface with wire junction installed.

Chapter 3

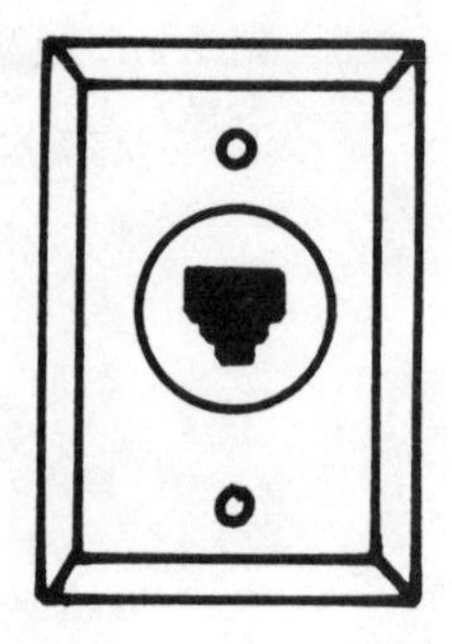

Apartment Installation

ALL OF THE WIRING IN NEW APARTMENTS IS RUN TO ONE central location (a closet or storage room) in the unit. (Older apartments may not be set up this way. Read *Extensions* at the end of this chapter.) In the newer apartments either the telephone company or the owner will put an interface next to all the wiring in that location. The telephone company supplies the dial tone to that interface. An outside apartment terminal looks like the one in Fig. 3-1. The serviceman will connect the station wire from this terminal to your apartment to provide the dial tone. The steps that follow show you how to take the dial tone from the interface to the jacks in the apartment.

TOOLS NEEDED

- ☐ Screwdriver
- ☐ Longnose pliers
- ☐ Wirestrippers
- ☐ Diagonal cutting pliers

EQUIPMENT NEEDED

- ☐ Jacks (flush)
- ☐ Wire junction, for the network interface, if needed

Fig. 3-1. Outside telephone terminal.

STEP-BY-STEP INSTALLATION (NEW APARTMENTS)

(1) Before you buy the jacks, check to see if the apartment complex will reimburse you for the jacks. This may save you an expense.

(2) Locate your phone jacks. Count them.

(3) Find out if a wire junction or jack (and short modular cord) is already connected with the interface or if you need to buy one. See Chapter 2 for directions on connecting it with the interface. (Note: There is one type of outlet that needs a special jack (buttons or snaps). Buy the correct type for the outlet.)

(4) See Chapter 8 to install. If there is any question about the colors of the wire in the apartment, see Chapter 7.

Note 1: Before you mount the jacks, check to see if the outlet box is the blue plastic type. These have a tendency to be hard to screw the jack bracket into. If you find this is the case, drill the holes larger with a 5/64-inch or 3/32-inch bit. The provided screws should then fit. An alternative to this, is to use one inch hex head screws instead of the provided screws. Screw them into the provided holes with a 1/4-inch nutdriver.

Note 2: There have been times when a wall jack will not hold a wall phone because the outlet was mounted too low next to a counter. If this happens, you can plug in a desk phone or with permission from the apartment manager mount the wall jack above the outlet.

(5) Install the wire junction (or jack) and connect all the apartment wiring to it. See *Rules and Regulations* (Chapter 2). Plug it into the interface. You should now have a dial tone at all of your jacks. If not, plug a phone into the interface. If you get a dial tone, there is something wrong in the wiring. See Chapter 11. If you do not get a dial tone, the trouble is with the telephone company.

EXTENSIONS

Most apartments are prewired so you usually cannot install extensions. If you need them, you could plug a long mounting or base cord into a working jack and hide the cord around furniture. There is even a tool (modular crimp tool) that you can buy to make custom length cords from bulk cord and modular plugs.

If you are set to run wire in a room, get permission from the manager, then follow the steps for running wire (Chapter 5). You cannot run wire outside the apartment, only inside.

Chapter 4

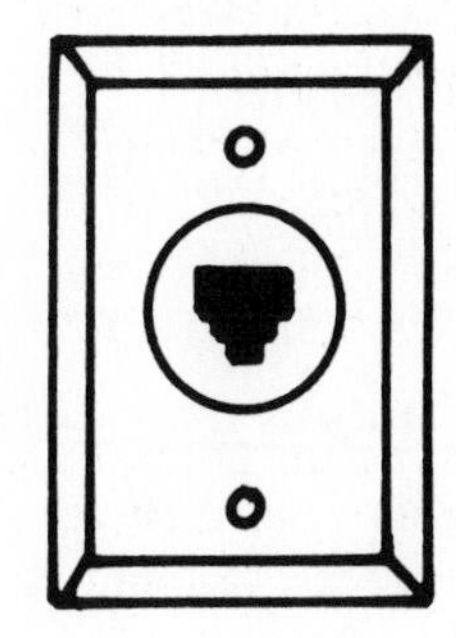

Homes Under Construction

THE DESIGNS OF MANY OF THE NEW HOUSES MAKE RUNning wire after the walls are up very difficult. It is important to plan the wiring before construction. Make sure that the house is wired for telephone service! I have seen houses finished with the utmost care but without a single length of telephone wire: a telephone installer's nightmare.

The best time to install phone wire is when the electrician is putting in his wire. This would be when the frame is up but the drywall has not yet been put in place. Sometimes the electrician will also install the phone wire because telephone outlets are placed next to power outlets.

More and more people independent to the electrician are being contracted to install telephone wiring. If you find that you want to do your own handy work, the directions that follow should make the prewiring of the home easy and inexpensive. Doing the task on the home yourself will bring joy.

You are responsible for the wiring from inside the home to the outside service connection. The phone company can place an outside interface. (See Interface in Chapter 2.)

Note: The following steps are best for the usual residence, but if you plan to install a multiline phone system to accommodate lights and hold, the wiring needs are different. Those instructions are located at the end of this step-by-step procedure.

PREWIRING OR ROUGHING IN

Tools Needed

- ☐ 3/8-inch power drill
- ☐ 1/2- to 5/8-inch wood bit
- ☐ Hammer
- ☐ Diagonal cutting pliers
- ☐ Tape measure

Equipment Needed

☐ Wire (see Chapter 7 for type desired)

☐ Outlet boxes (I suggest the type that have two nails so that all you need to do is nail directly to the stud.)

☐ Romex staples

Step-by-Step

(1) Make a diagram or use the blueprints of the home to locate the most convenient places for the jacks.

> ☐ You may want to discuss this with your builder or electrician.
>
> ☐ Telephone outlets will usually be located across the room from the TV cable. (The only reason not to do this is if the television has a telephone answering device.)
>
> ☐ Telephone outlets should first be placed in the master bedroom, kitchen, and living room. The next are the other bedrooms, playrooms, garage, and any other areas connected to the home where you will desire service later.
>
> ☐ If phone outlets are placed close to corners the movement of furnishings may be more flexible.
>
> ☐ Try to place telephone outlets near electrical outlets for convenience to furniture and use of phone devices (such as modems and answering machines).

(2) Plan the route of the wire and draw it on the diagram. Make the route as direct as possible. Right angles can be very hard to pull. You will either want to go through the ceiling joists or the basement joists (whichever is easier in your home).

> ☐ There are three types of routes you can take in running wire. See Chapter 7.

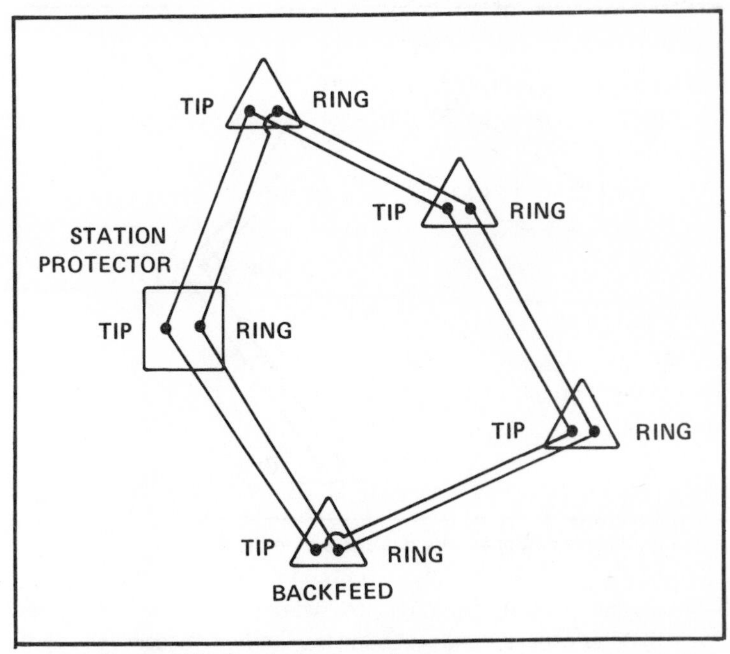

Fig. 4-1. How backfeeding looks on a wire run.

—(Route 1) a complete circle: one wire runs through all the jacks back to the starting point. This is called back feeding. See Fig. 4-1.
—Advantage—If wire is cut there is an alternate route to supply the dial tone.
—Disadvantage—A wire run this way is sometimes difficult to pull. It may be easier (if the distance is very great) to run two separate wires from the interface to the farthest outlet and connect them in the jack.

—(Route 2) starts at one point and ends with the last jack. See Fig. 4-2.
—Advantage—will pull from one point to the last jack fairly easily.
—Disadvantage—If wire is cut or damaged all jacks from that point on will not work.

□ Both wires will be run to the power meter where the outside service (interface, etc.) will be connected.

(Route 3) each jack has its own wire back to the point of demarcation. This is called a "home run." This method should be used in a large home that contains more than 10 jacks.
—Advantage—It is easy to locate trouble.

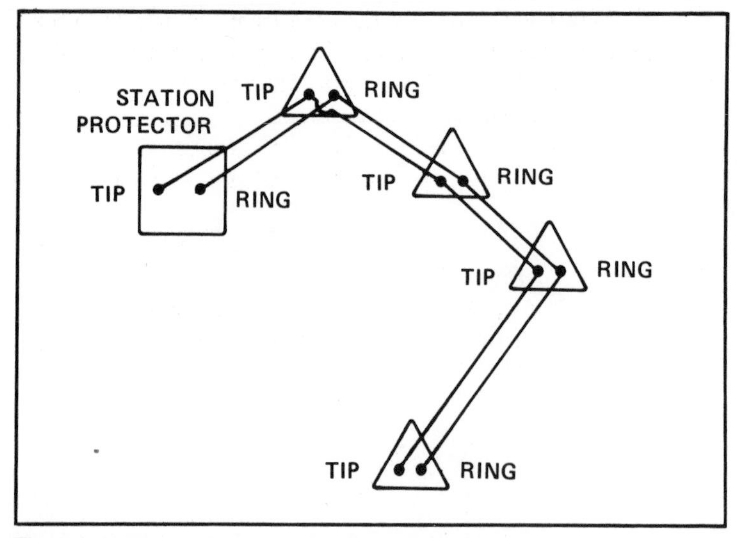

Fig. 4-2. Multiple or looped wire run one end to the other.

—Disadvantage—It takes longer to install.

(3) After choosing the route and the method, measure the wire to all of the outlets.

☐ Measure the vertical distance from the power meter to the ceiling joist or basement joist.

Fig. 4-3. Outlet box mounted to stud at measured height.

☐ Measure the vertical distance from the outlets to the joists (eight feet if standard ceiling or two feet through the basement). Double the length (the wire will be looped through the outlets and back).
☐ Measure the route, at a convenient level, to each outlet.
☐ Add the three measurements together (remember about doubling).
☐ Multiply the total distance by one and one half (to allow for error and slack).
☐ If running quad, run two wires at the same time (so that later you will have a spare). Multiply the total distance by two for the two wires to be pulled.

(4) Buy equipment.

☐ Wire
☐ Outlet boxes

(5) Install the outlet boxes at the same height as the power outlet boxes so that they will look neat. See Fig. 4-3. The kitchen phone will usually have the outlet box 50 to 52 inches above the floor. Consider the area carefully so that the cord does not cross over a stove or end up behind a refrigerator or other appliance. Since this outlet will be for a wall phone, be sure to leave enough space around it for the phone itself.

(6) With the power drill and bit, drill the holes through which the wire will be routed. These holes will probably be in the top or sole plates and joists of the walls and ceilings. Figure 4-4 shows

Fig. 4-4. Drilling through top plate for wire run.

Fig. 4-5. Romex staples in stud for running wire.

the drill going through the top plate. Make sure the wire will not be in the way when the drywall is erected.

(7) You may want to place Romex staples on the studs to run wire down to the outlet if it is coming from the top plate. About six are plenty. This will keep the drywall installation from harming the wires. See Figs. 4-5 and 4-6.

(8) Pull the wire through the holes and past each outlet consecutively.

(9) The end of the wire should be at the farthest outlet or back to the interface wire junction.

(10) If there are two quads, tag the wire that is going to have the dial tone first. (This should be done at each outlet and outside).

(11) Pop the knockout from the outlet box.

Fig. 4-6. Wire run through romex staples to outlet box.

Fig. 4-7. Looped or multiple uncut station wire in outlet box.

(12) Pull a loop of the wire into each outlet leaving about 8-inches slack.

(13) Loop the wire into each outlet and tie it as shown in Figs. 4-7 and 4-8.

(14) Drill a hole less than 12-inches below the power meter.

(15) Pull the wire through to the outside.

(16) Seal the hole around the wire with silicone caulk to keep out moisture and insects.

(17) The telephone company will now give you the dial tone outside.

(18) Save your diagram locating the outlets so they can be found in case one gets covered when the walls are put up.

(19) Install jacks.

INSTALLING THE JACKS

Tools Needed

- ☐ Screwdriver
- ☐ Diagonal cutting pliers
- ☐ Longnose pliers
- ☐ Wire strippers

Equipment Needed

- ☐ Jacks (flush type)
- ☐ Wire junction (for interface connection if applicable)

For a description of the jacks that you might put in, see Chapter 8. The wire junction will go in the same location as the inside interface. When you receive the dial tone from the telephone company, plug the modular cord from the wire junction into the interface, or terminate to the wiring bridge of the outside interface. This will give service to all the jacks.

MULTI-LINE SYSTEM

If you decide to have a multi-line system, installation will be like that earlier in this chapter except that: (1) the wire will be routed differently (home run), (2) you will need a closet with a dedicated electrical outlet to run the main telephone switching cabinet for the electronic phones, (3) the telephone company will need to install interfaces in the closet, and (4) you must number each station wire, when you pull it, to indicate which jack it runs to.

Each station wire (at least four-conductor, eight-conductor is safer) will be routed back separately from a jack to this closet and terminated on a termination block. The block will have split-beam terminations for all the wires to be connected.

Fig. 4-8. Looped or multiple run station wire cut in outlet box later to be terminated in jack.

Chapter 5

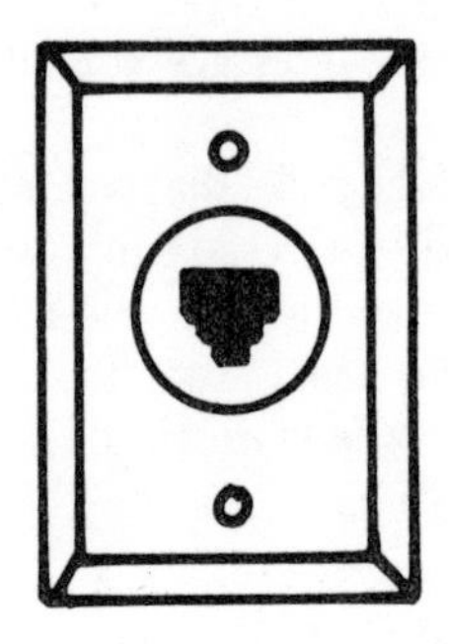

Homes with Existing Service

IF YOU HAVE PHONE SERVICE ALREADY ESTABLISHED IN YOUR home and you want to add an extension in a convenient place, use this section. **Caution:** do not work without telling someone what you are doing in case there is an emergency. Be sure the area you drill through is free of electrical wires and other hazards. Do not work in an attic when the temperature is hot.

Tools Needed

☐ See specific method of installation

Equipment Needed

☐ See specific method of installation

Step-by-Step Installation

(1) Decide where you want to install the new jacks.

(2) Make a diagram of the home showing where all the working jacks are presently located.

(3) Locate where the new jacks will be on the diagram.

(4) Decide if the most accessible path to run wire is to an existing jack, a wire junction, or the wire junction adjacent to the interface.

(5) Choose the method of installation best suited to your home:

☐ Wire run from one room to another.
☐ Wire run through the crawl space or basement under the home.
☐ Wire run through the crawl space or basement, then inside the wall.
☐ Wire run around the outside of the house.
☐ Wire run through the attic.
☐ Wire run through the attic, then inside the wall (dropping a wall).

WIRE RUN FROM ONE ROOM TO ANOTHER

Tools Needed

☐ 3/8-inch power drill
☐ Bell Hanger Drill 1/4-inch diameter and 12 or 18 inches long.
☐ Screwdriver
☐ Longnose Pliers
☐ Wirestripper
☐ Diagonal cutting pliers
☐ Staple gun or insulated staples and a tack hammer
☐ 1/4-inch nut driver and hex head screws or (push drill with No. 30 bit optional)

Equipment Needed

☐ Wire (four conductor)
☐ Jack (nonflush or surface jack on the baseboard)
☐ #8-3/4 inch or 1-inch pan head screws

Step-by-Step Installation

(1) Choose a visually unobtrusive path for the wire (Fig. 5-1).
(2) If necessary, drill through the wall with a bell hanger drill.
(3) To run wire

☐ Around the edge of the carpet—hide the wire between the carpet tackless strip and the wall (Fig. 5-2). You may have to pull up a section of carpet at a time with the longnose pliers if the wire can not be slid into the area. (Caution: be careful not to cut your fingers on the tackless strip).

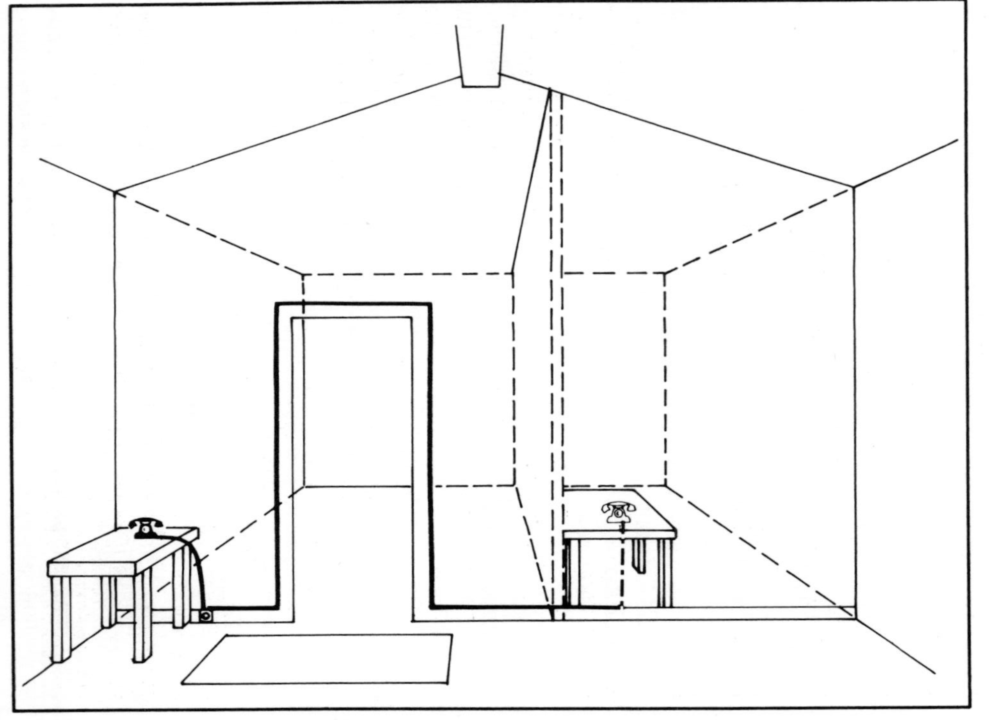

Fig. 5-1. Station wire run from room to room.

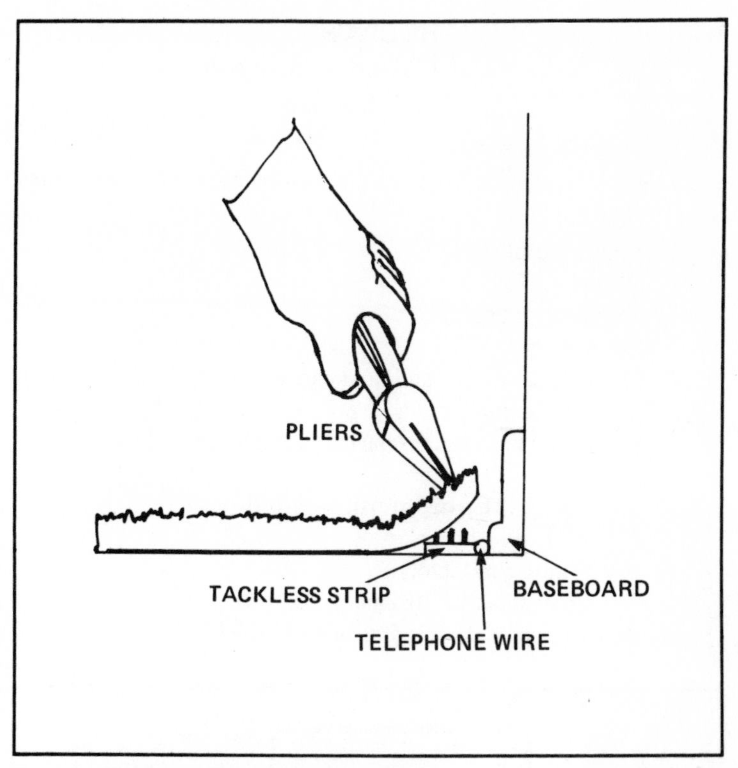

Fig. 5-2. Wire run around the edge of carpet.

☐ Wood molding—Staple along the top of the baseboard molding. To staple around the door trim, double staple 1 inch from the door trim. Make a 90 degree bend. Double staple 1 inch from the bend on the door trim. Pull the slack taunt over the top of the door trim to form a straight line. Double staple 1 inch from the top corner. Staple all the way to the bottom spacing 7 1/2 inches or the length of the staple gun from each staple. Always double staple at each corner (Fig. 5-3). (Note: If possible, avoid stapling on the drywall since it will not grip staples properly).

(4) When you reach the working jack to which you will terminate, *do not* staple next to it until the outer sheath is stripped and terminated on the jack. This is also true for the new jack you are installing.

(5) See Chapter 8 for how to terminate connecting block and jack.

WIRE RUN THROUGH THE CRAWL SPACE OR BASEMENT UNDER THE HOME

Tools Needed

☐ 3/8-inch power drill

☐ Bell Hanger Drill 1/4-inch in diameter and 12 or 18 inches long

☐ Longnose pliers

☐ Wirestrippers

☐ Diagonal pliers

☐ Screwdriver

☐ 1/4-inch nut driver and hexhead screws or (push drill with a No. 30 bit is optional)

☐ Staple gun (optional) or insulated staples and a tack hammer

Equipment Needed

☐ Wire (four conductor)

☐ Jack (non-flush or surface jack)

☐ #8 3/4-inch or 1-inch pan head screws

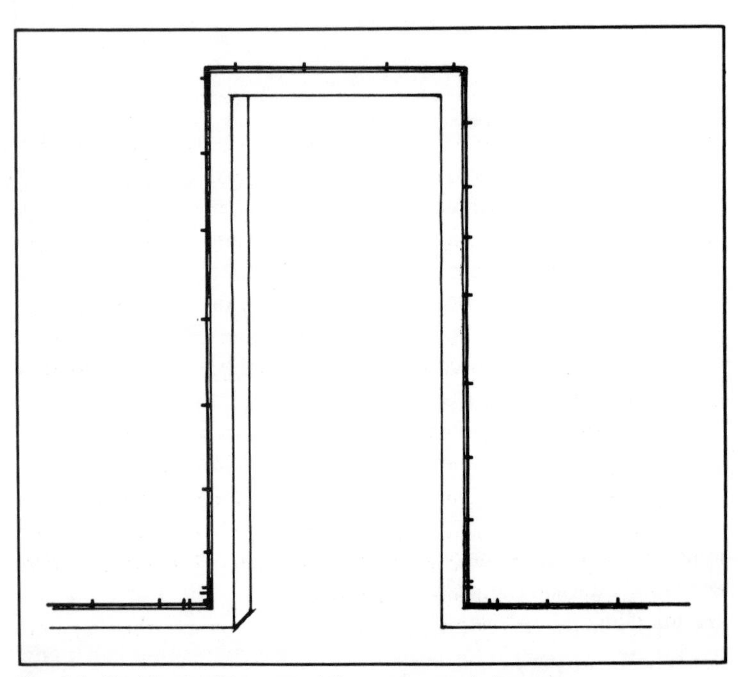

Fig. 5-3. Double stapling around the corners and door trim.

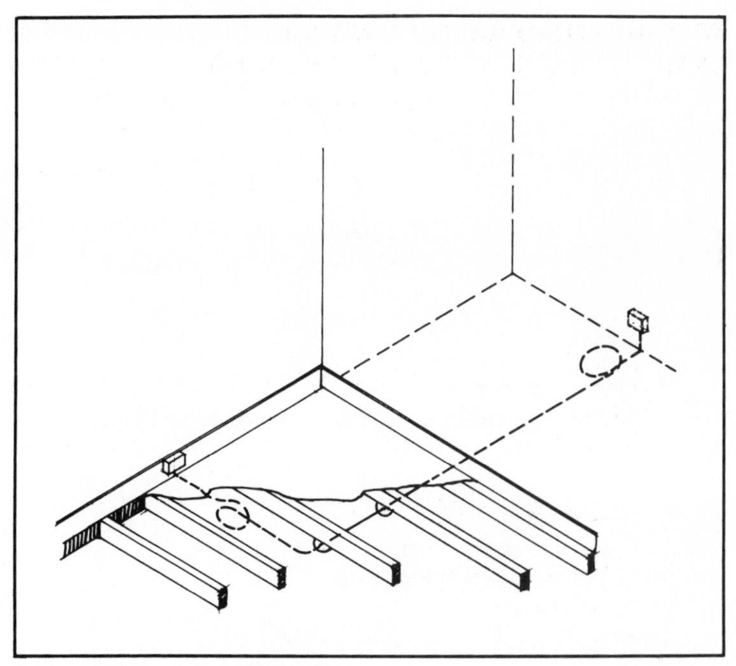

Fig. 5-4. Wire run under the home.

Step-by-Step Installation

(1) Drill a 1/4-inch hole through the floor where you want the extension jack to be placed (Fig. 5-4). Drill another hole through the floor where the working jack will be connected. (Note: When drilling a hole next to carpet, be sure to pull the carpet back enough that the drill does not catch the pile and tear it.)

(2) Push the new wire through one hole for easy access underneath.

(3) Push a small piece of spare wire through the other hole to make it easy to find.

(4) Pick up the wire (that you are going to pull) from under the crawl space.

(5) Run it through the beams under the home. Keep it off of the ground to prevent problems later. (Note: *Do not* staple the wire. Run it through rings or Romex staples. This will make the wire easy to pull later if you need to repull.)

(6) Go to the hole where the other tracer wire is showing and pull it out.

(7) Push the wire up through the hole. It may help if someone

is on top to grab the wire and tie a knot in it to keep it from falling back.

(8) Optional—Staple each wire above the hole to keep it from falling back while you are working on it

(9) See Chapter 8 for how to terminate jack.

WIRE RUN THROUGH THE CRAWL SPACE OR BASEMENT THEN INSIDE THE WALL

Tools Needed

- ☐ 3/8-inch power drill
- ☐ Bell Hanger Drill 1/4-inch diameter and 12 or 18 inches long
- ☐ Screwdriver
- ☐ Longnose pliers
- ☐ Wirestripper
- ☐ Diagonal cutting pliers
- ☐ Hole cutter or keyhole saw
- ☐ Coathanger
- ☐ Folding ruler or yardstick
- ☐ Utility light, if necessary

Equipment Needed

- ☐ Wire (four conductor)
- ☐ Jack (flush)
- ☐ Wall anchors
- ☐ Outlet box (Depends on how jack will be mounted)

Step-by-Step Installation

(1) Drill a 1/4-inch hole through the floor where you intend to mount your flush jack so that you can find the sole plate easily when you are beneath the home (Fig. 5-5). Push a 3-foot piece of wire into the hole for you to find.

(2) Go beneath the home, find the trace wire, and drill up into the hollow wall through the sole plate.

(3) Pull the wire, from the splicing location, over the beams or through rings.

(4) Push the end of the wire into the hole you drilled.

(5) Go to where you plan to install the jack and measure the height of the other outlets so this jack will match.

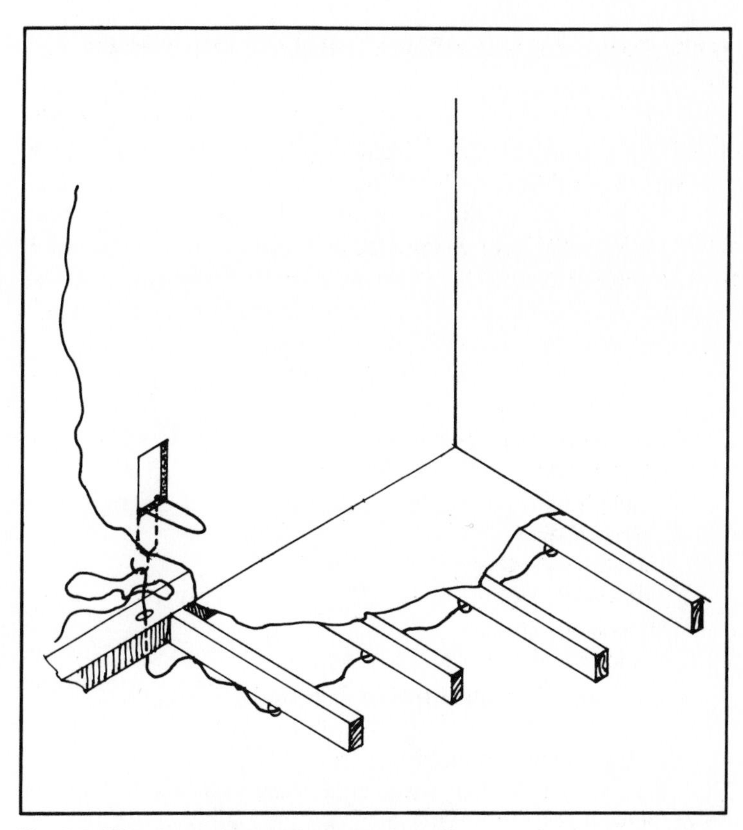

Fig. 5-5. Wire run under crawl space into wall.

(6) Cut the hole for your flush jack with either the hole cutter or keyhole saw. (Note: Check to the right and left of where you plan to cut the hole by pushing a screwdriver through. This is to see that you do not hit a stud in the wall.)

(7) Use the coat hanger you have made into a hook and fish the wire out.

(8) Mount the jack.

- ☐ If you are just going to mount the jack directly to drywall use wall anchors.
- ☐ If you choose to mount the jack in an outlet, check Equipment needed. Pull the wire through the outlet before you mount the outlet.

(9) See Chapter 8 for how to terminate jack at both ends.

WIRE RUN AROUND THE OUTSIDE OF THE HOME

If the outside of your home is masonry or brick, you may want to avoid running outside the home (Fig. 5-6). If you decide to do it anyway, you will need, in addition to the tools shown, a 12-inch masonry drill and wire fasteners for masonry. You can use carbon black cable ties on conduit if you use it for part of the wire run.

Drill part way through the wall with the "bell hanger drill for wood" until it reaches the masonry, then finish the hole with the masonry drill. If you have a bell hanger masonry drill you can pull the wire back through the wall.

Tools Needed

- ☐ 3/8-inch power drill
- ☐ Bell Hanger Drill 1/4-inch diameter and 12 or 18 inches long
- ☐ Staple gun or insulated staples and a tack hammer
- ☐ Screwdriver
- ☐ Longnose pliers
- ☐ Wirestrippers
- ☐ Diagonal pliers

Equipment Needed

- ☐ Wire (Four conductor)
- ☐ Jack—non-flush or surface. (A flush jack may be used on an outer wall.)
- ☐ #8 3/4-inch or 1-inch pan-head screws
- ☐ Silicone caulk
- ☐ Wall anchors (On flush jacks)

Step-by-Step Installation

(1) Decide the best route to run the wire around your home. (See Chapter 7.) Splice the new wire to a working jack, the protector, or a wire junction.

(2) Drill through to the outside using a bell hanger drill. Drill at a slight downward angle to help prevent moisture from coming in. Drill where the new jack is going and also where the inside termination will be made, if needed.

(3) Pull the wire outside through one hole.

(4) Take the wire around the home and bring it back to an inside jack or to the protector, if applicable. The wire can be pulled in with the bell hanger drill by using the hole in its shank.

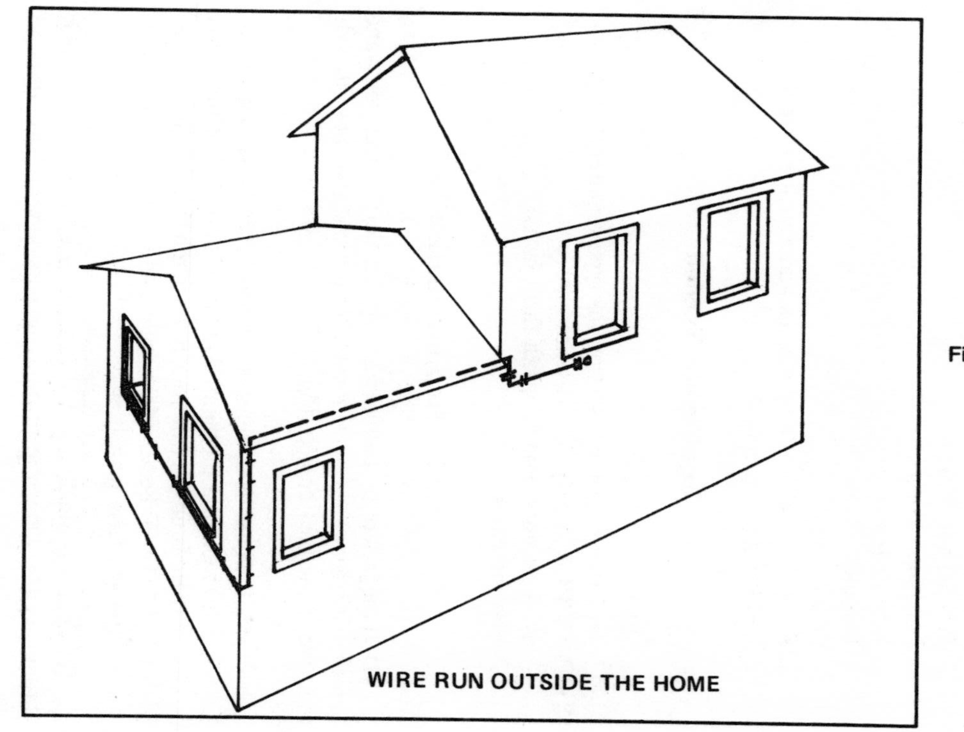

Fig. 5-6. Wire run outside the home.

(5) Staple every 7-1/2 inches and at 90 degree angles double staple 1-inch from each bend.

(6) At the hole where the wire goes in make a drip loop to help to keep water out.

(7) Put silicone caulk in the holes to seal out moisture and insects.

(8) See Chapter 8 on termination of the jack you selected.

WIRE RUN THROUGH THE ATTIC

Tools Needed

- ☐ 3/8-inch power drill
- ☐ Bell hanger drill
- ☐ Longnose pliers
- ☐ Wirestrippers
- ☐ Diagonal Pliers
- ☐ Screwdriver
- ☐ Staple gun or insulated staples and a tack hammer
- ☐ Utility light, if necessary

Equipment Needed

- ☐ Wire
- ☐ Jack (see Chapter 8 section for type required)
- ☐ #8 3/4-inch or 1-inch pan-head screws
- ☐ Wall anchors (For use with flush jacks)
- ☐ Silicone caulk

Step-by-Step Installation

(1) If the jack or protector (when applicable) to which you want to terminate, is on the opposite side of the home, the path through the attic is the best choice (Fig. 5-7).

> ☐ It would be better if you can walk around the attic without much of a problem. Always walk on top of the joists if there is no floor or your foot will go through the ceiling.
>
> ☐ Wear a dust mask, goggles, and a long sleeve shirt if the attic has insulation.

(2) Drill a 1/4-inch hole and run the wire from an inside jack through a closet ceiling or through the outside wall. (Drill through

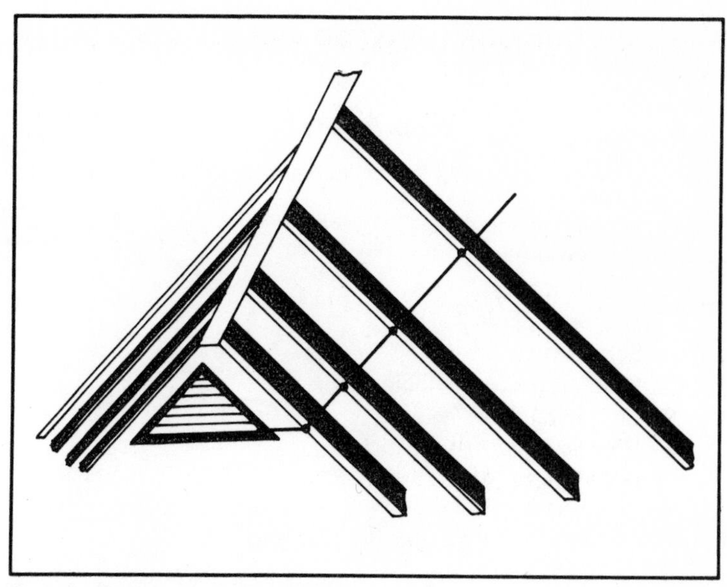

Fig. 5-7. Wire run through the attic.

a gable or soffit (eave) vent into the attic, preferably in a corner where it will not be noticeable).

(3) Run the wire over the beams in the attic to keep them out of the way. (Put the wire in rings or Romex staples so it will not be cut or mashed if something is stacked on it).

(4) Drill a hole to the location of the new jack. This can be through the outside corner of a closet or through the outside wall. Pull enough slack to staple along the baseboard or trim to the new jack.

(5) If you drill through the outside wall:

- ☐ Drill through a gable or soffit (eave) vent.
- ☐ Run the wire outside to the new location.
- ☐ Drill at the new location, from the inside outward with a downward slope).

- ☐ Staple the wire to the outside wall at 7-1/2 inch spacings on wooden trim.
- ☐ Make drip loops to keep water out.
- ☐ Seal all outside wall holes with silicone caulk.

(6) See Chapter 8 for termination of jacks.

WIRE RUN THROUGH THE ATTIC AND THEN INSIDE THE WALL (DROPPING A WALL)

Tools Needed

- ☐ 3/8-power drill
- ☐ 1/2-inch spade drill
- ☐ Longnose pliers
- ☐ Wirestrippers
- ☐ Diagonal pliers
- ☐ Screwdriver
- ☐ Hole cutter or key hole saw
- ☐ Coat hanger for hook
- ☐ Folding ruler or yardstick
- ☐ Utility light (if necessary)

Equipment Needed

- ☐ Wire (four conductor)
- ☐ Jack (flush jack)
- ☐ Wall anchors
- ☐ Electrician's vinyl tape (black)
- ☐ Outlet box (optional)

Step-by-Step Installation

This is the method that will look the best but it is also the hardest and sometimes is not even possible (Fig. 5-8). Read the steps and be sure the walls are not obstructed (fire walls or fiberglass insulation). If so, you will not be able to drop walls easily. Outside walls are usually insulated. Be sure to walk on the joists. Wear a dust mask, goggles,and a long sleeve shirt if the attic is insulated.

(1) Decide where you want the jack to go and mark the ceiling directly above by making a small hole with a small screwdriver. Leave it in the ceiling so that you can find it when you are in the attic.

(2) Drill through the top plate of the wall with a 1/2-inch spade drill above where you intend to place the new jack. Refer to Fig. 4-4.

(3) Drop a length of wire or string with a weight on the end to be sure that there are no obstructions. If there is a firewall it will usually be 48 inches above the floor.

(4) Terminate the new wire to the working jack by using the

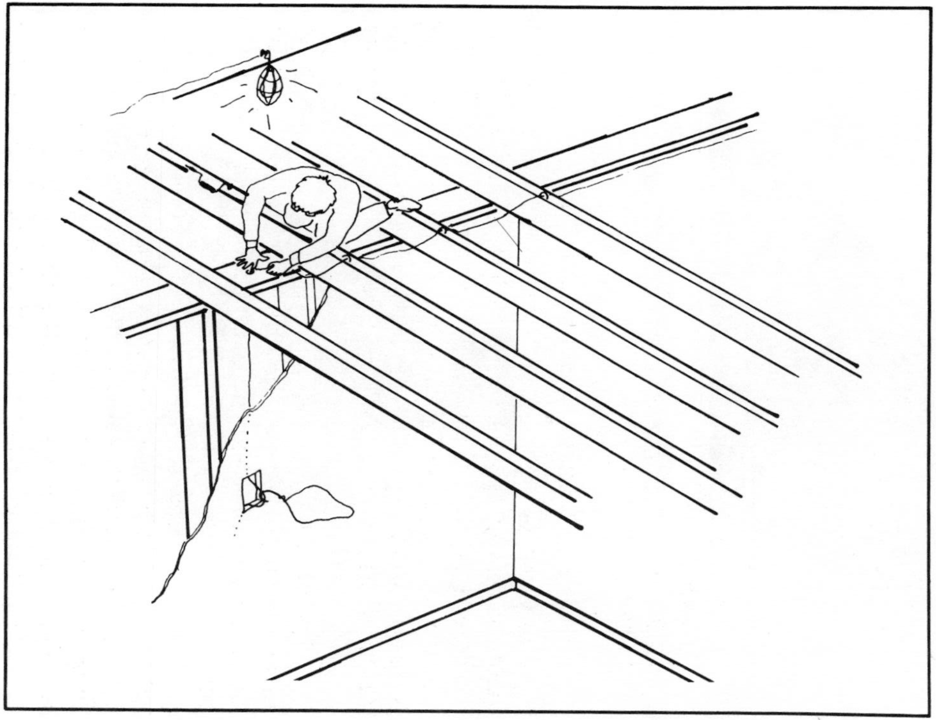

Fig. 5-8. Dropping wire inside the wall from the attic.

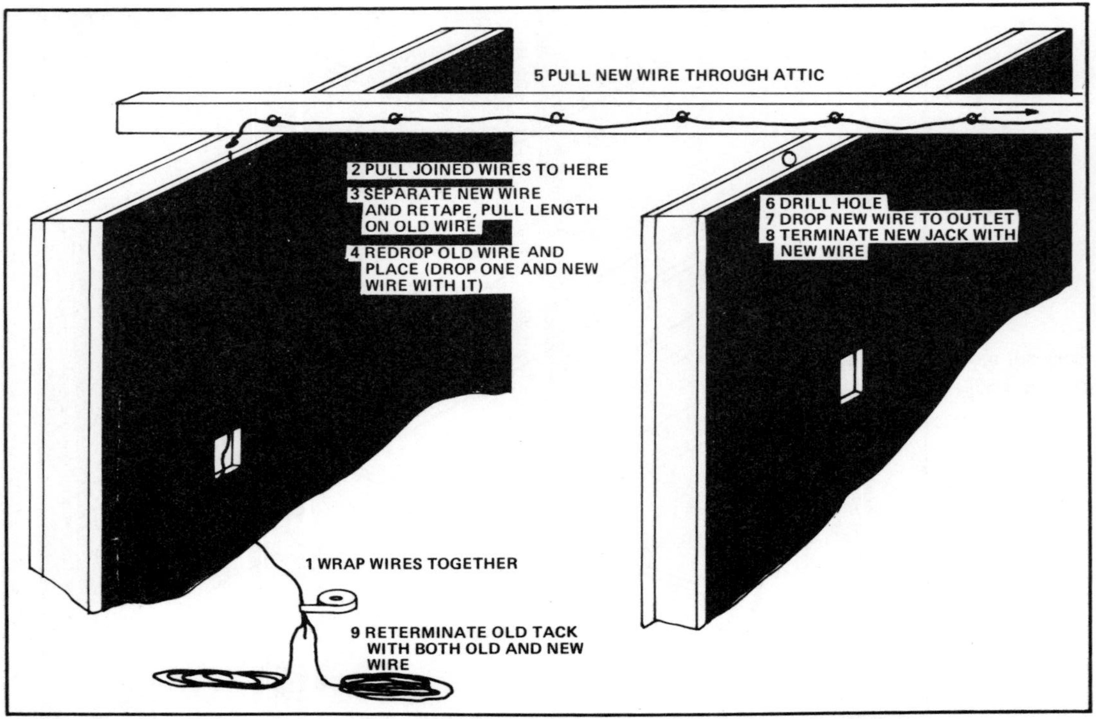

Fig. 5-9. How the existing wire is used as a pull string.

existing wire as a pull string (Fig. 5-9). (See Chapter 7 for how to tie wires to be used as a pull string)

(5) Check to see which side of the stud the wire will drop on. Make a hole for the jack in the wall at the same height as other wall outlets by measuring with a ruler (Fig. 5-10). The hole should

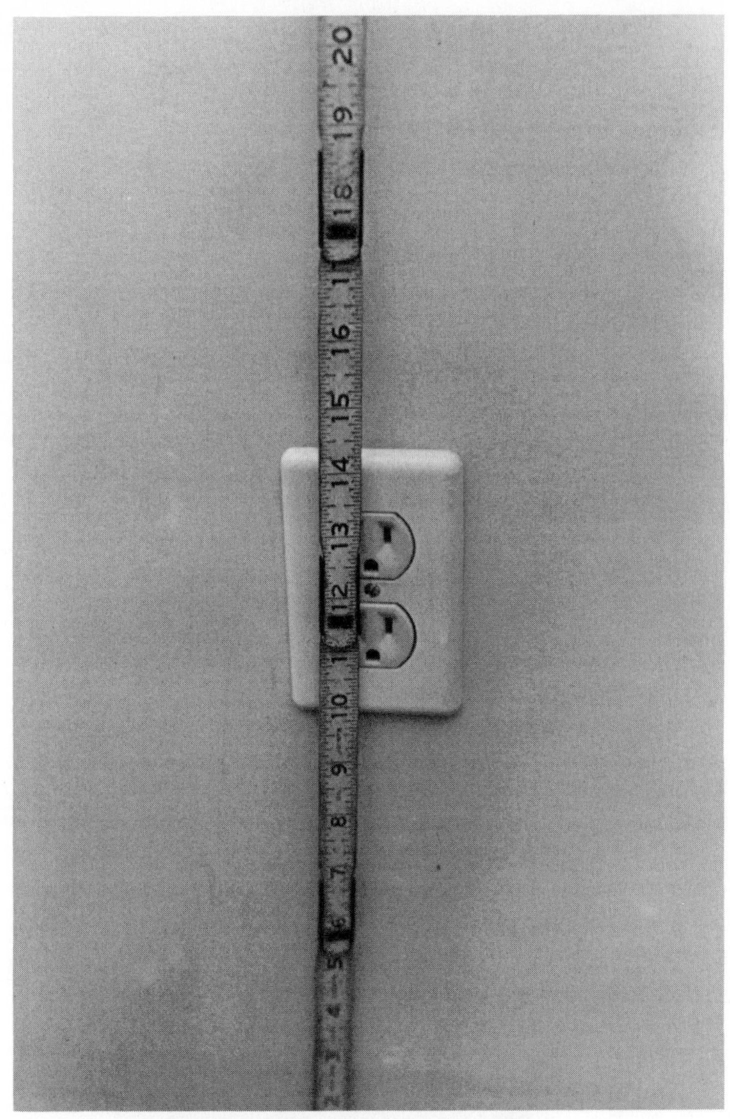

Fig. 5-10. Measuring the height of an outlet with a ruler.

Fig. 5-11. Hole cut in drywall with hole saw at proper height.

not be any larger than 1-3/4 inches in diameter. If you are mounting a rectangular flush jack directly on drywall, use a 2-1/4 inch hole saw (Fig. 5-11).

(6) Drop the wire by itself or tied to a weighted string.

(7) Fish out the wire or string with bent coathanger.

Fig. 5-12. Plumb bob bracket into alignment.

Fig. 5-13. Fastened bracket with Molly® bolts to wall.

Fig. 5-14. Terminate jack.

(8) Terminate new wire on the working jack to which you are going to bridge. See Chapter 8.

(9) Use a plumb bob to position the bracket. See Fig. 5-12.

(10) Mark where you want the wall anchors.

(11) Attach the wall anchors.

(12) Fasten the bracket of the new flush jack to these wall Molly® bolts. Fig. 5-13.

(13) Terminate new jack. Figure 5-14 shows one jack terminated. See Chapter 8 for details.

Chapter 6

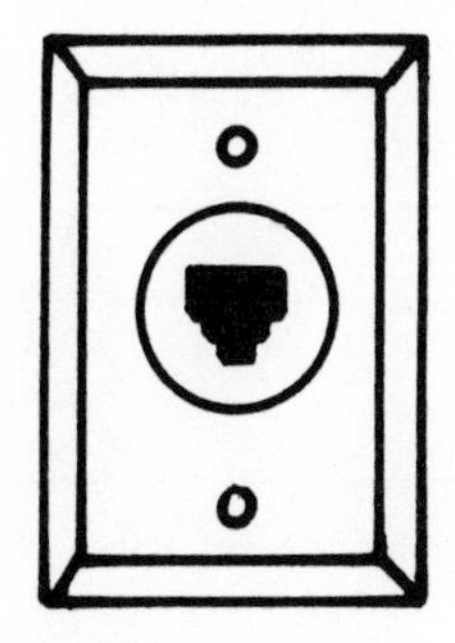

Mobile Homes and Portable Buildings

INSTALLING JACKS IN A MOBILE HOME IS EASY SINCE A MObile home is not built at all like other dwellings. Recreational vehicles are not mobile homes but they can be wired for phones by the phone company with special fast connect-disconnect outlets. Safety tip—Have the electric company check to be certain the mobile home is grounded. (A mobile home is metal and can become a live wire.)

TOOLS NEEDED

- ☐ Screwdriver
- ☐ Longnose pliers
- ☐ Wirestrippers
- ☐ Diagonal cutting pliers
- ☐ 3/8-inch power drill
- ☐ Bell hanger drill
- ☐ Hammer (for Romex staples)
- ☐ Staple gun (optional)

EQUIPMENT NEEDED

- ☐ Wire (four conductor)
- ☐ Jacks (non-flush or surface) Note: You can install a wall-

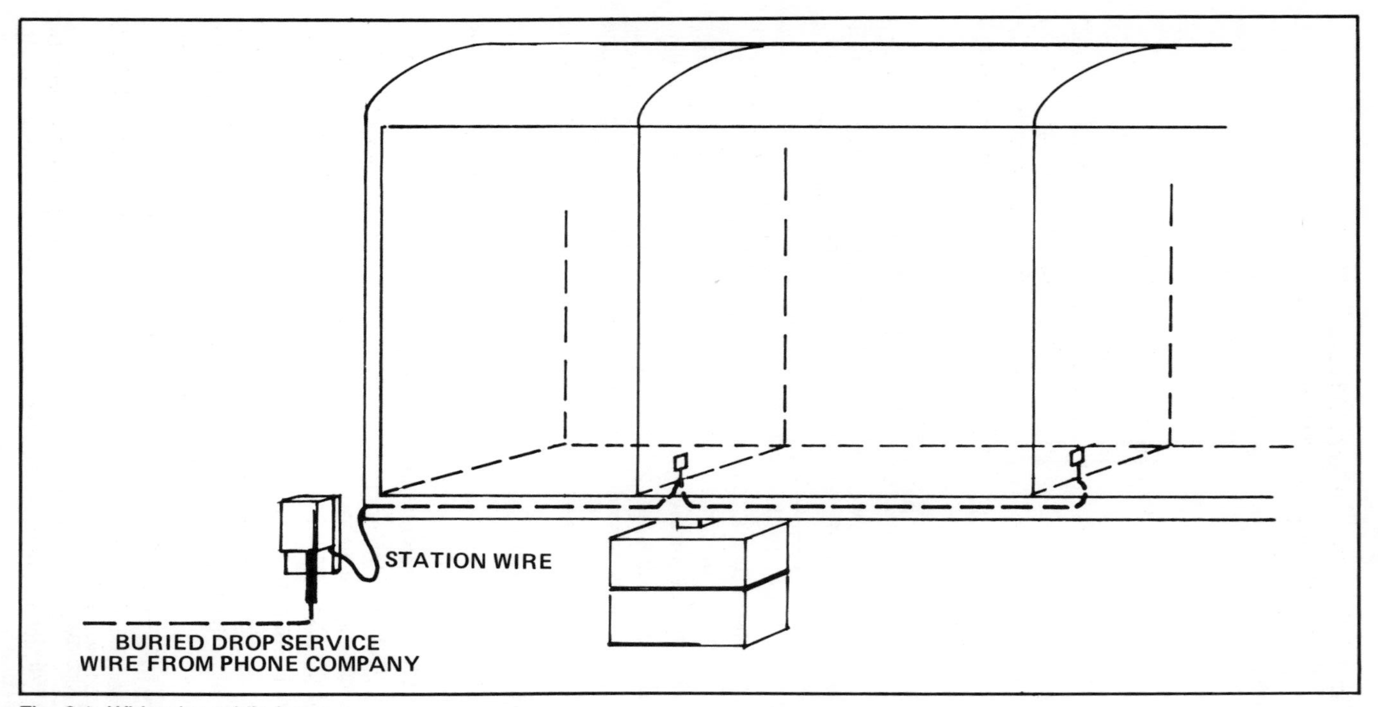

Fig. 6-1. Wiring in mobile home.

phone jack without an outlet box.

- ☐ #8 3/4- to 1-inch pan head screws
- ☐ Duct tape
- ☐ Romex staples

Step-by-Step Installation

(1) Do not drill through the side of the mobile home. This would leave the mobile home open to water and other problems. Also, there is no way to fasten wire properly to the sides of a mobile home. Plan your route. Figure 6-1 shows the most common way to run wire in a mobile home.

- ☐ Decide where you want the jacks. You may need to place them close to an outside wall since the main services are usually run the length of the mobile home through the middle of the floor.
- ☐ Plan the route so that only one station wire is run to the outside protector or interface. (Either run one continuous wire or join them at a jack or wire junction prior to running outside.) The phone company will install the interface when they give you service. They will not mount it directly on the mobile home, but on a post or stake nearby. Leave enough slack in the wire to reach comfortably.

(2) Drill down, where you want each jack, through the plywood floor. Avoid the outriggers (metal supports).

(3) Push the wire down through the hole you drilled in the floor.

(4) Find the spots under the mobile home where the drill went through. At each place, cut a small slit with a knife. Grab the wire and pull it out. An alternative method is to drill through the floor and pull the wire up with a bell hanger bit.

(5) Run wire under the home through the openings in the framework or run the wire on the outer edge of the outrigger to its termination at the protector or outside interface. You may staple the wire to the outer edge if the material is solid.

(6) Seal the slits you made under the mobile home with duct tape to keep the air and moisture out.

(7) When connecting the jacks to the wall try to terminate on a stud since the thin materials of the wall will not hold jacks securely

as they should. The studs are usually 16-inches apart at the center of the outrigger area. You can find the studs by "sounding." (Hit the wall and listen. A solid sound is a stud and a hollow sound is in between the studs). See Chapter 8 for bridging and termination).

Chapter 7

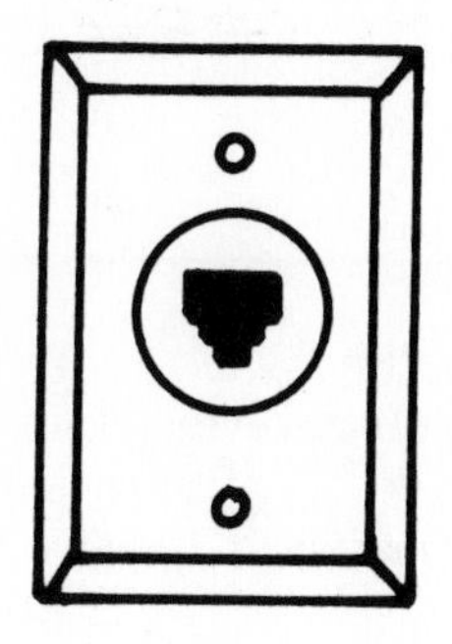

Wire

THE THREE CATEGORIES OF WIRE FOR USE IN THE HOME are four (or three) conductor, multiconductor, and odd types. The wires create a simple electrical circuit with a battery side and a ground side. The battery side is called "ring" and the ground side is called "tip." These terms originated with the plugs of the old style manual switchboards. The point was called the "tip" while the sleeve adjacent to it was called the "ring." The wires are color coded for easy identification. The green wire is usually the tip of the circuit and the red wire is the ring. The black and yellow wires are a square pair. They are used for extra functions such as a secondary telephone line or number, or supplying power from a transformer to a light in a telephone dial or to a buzzer for signaling. See Fig. 7-1.

Wire is found in different gauges (AWG) from #19 (large) to #26 (small). The #22 to #24 gauges are the easiest to terminate on telephone jacks. When you are considering the various gauges, remember that the smaller the gauge the more resistance on the wire. Do not use a small gauge wire if you are going to have several phones on one line. The best gauge for residence use is #22.

FOUR CONDUCTOR

(Two pair, Quad, or Station wire)

☐ (1) Green, Red, Black, and Yellow

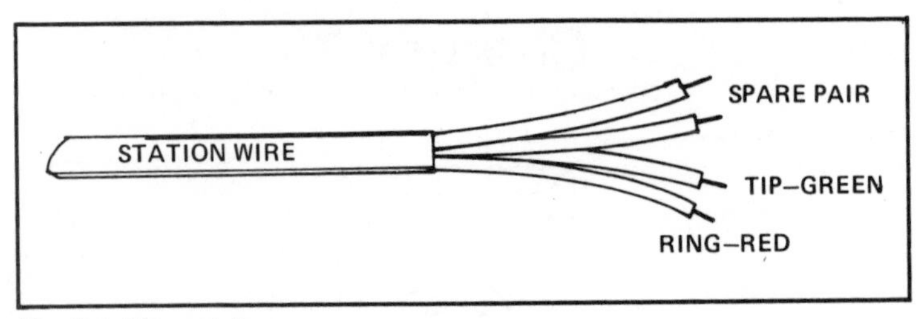

Fig. 7-1. Tip and ring.

□ (2) (White-blue, Blue-white) (White-orange, Orange-white)

Three conductor wire is:

□ Green, Red, and Yellow

□ Quad or (four conductor) is the most common telephone wire. It has the four color-coded conductors; green, red, black, and yellow. The green and red wires supply the dial tone to the jack. The earlier wires of this type are either rubber or cloth insulated. The rubber insulated wires will have colored threads beneath the insulation against the copper wire. The cloth type may have a colored band around the outside. Older wire was fastened, with the aid of tacks, to baseboard or trim. When you remove an old jack and connecting block, label the wires to indicate tip and ring (attached to phone cord spade tip leads).

□ On some four conductor wire the color code is not green, red, black, and yellow but white-blue, blue-white, white-orange, and orange-white. This should be treated the same as the g,r,b,y color coding with the white-blue wire replacing the green wire and the blue-white replacing the red wire. The white-orange and orange-white replace the black and yellow of the second pair.

□ The three conductor type has the green, red, and yellow insulated wires but is missing the black. It should only be used to run an extension.

MULTICONDUCTOR

The expansion of telecommunications has increased the number of families who need two or three different numbers in their home. The single four conductor wire cannot handle more than two lines, so when an additional line is needed more wire must be run to the location. Prewiring with multiconductor wires is easier than running two or more four-conductor telephone wires. Also all the conductors are color coded. Multiconductor wire is often, however, a small gauge wire (usually 26) so you will need to consider how many phones you need on the line.

The color code for multiconductor wire begins wb,bw,wo,ow . . . like four-conductor wire. The color code for all multiconductor wire is compiled from ten main colors. The tip color codes of the pair are white, red, black, yellow or violet. The ring color codes of the pair are blue, orange, green, brown, or slate. The dominant color on an insulated wire is always listed first and the band color is listed second (example: a red wire with a blue band (trace color) is called red-blue and is the tip wire of the red-blue and blue-red pair of wires). Some paired wires may not have the band or trace color. They only have the solid colors of the color code for tip or ring identification. You must be sure not to separate the paired wires from one another or you may end up having to trace the mate. Multi-conductor wire has a nylon string inside it with the colored conductors (Fig. 7-2). This string is used to peel the sheath off the wires easily.

Fig. 7-2. Color coded wire.

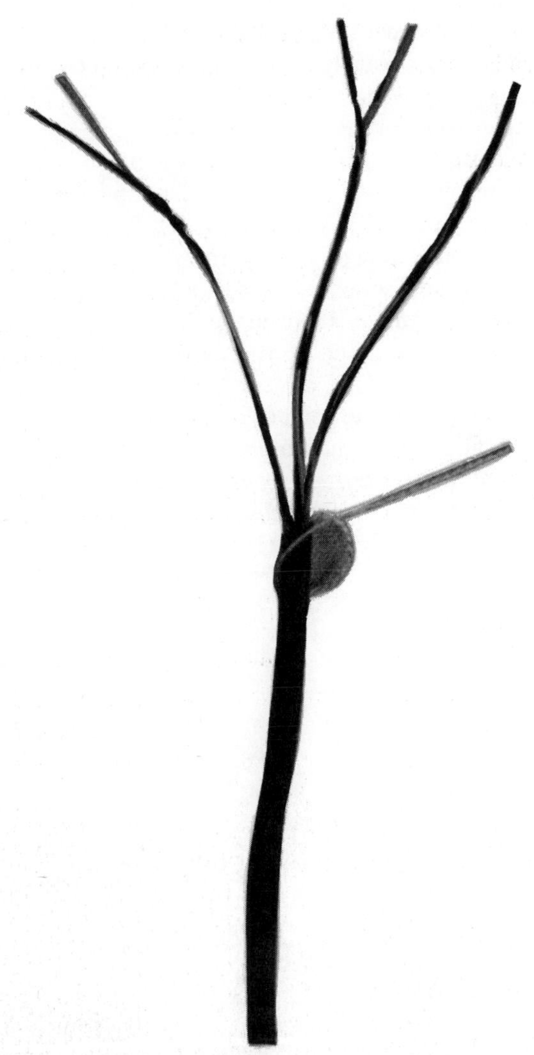

Fig. 7-3. Multiconductor wire with string for stripping sheath.

SIX, EIGHT, AND TWELVE CONDUCTOR

Three Pair

(1) White-blue, blue-white; white-orange, orange-white; white-green, green-white. See Fig. 7-3.

(2) Green, Red, Black, Yellow, White, and Blue

Four Pair

White-blue, Blue-white; White-orange, Orange-white; White-green, Green-white; White-Brown, Brown-white. See Fig. 7-4.

Six Conductor

(1) Three-pair telephone wire is the same as four conductor except that it has two extra conductors (or a pair). The color code on this pair is different. (See (2) of four-conductor wire). The last pair is the White-Green, Green-White. This pair makes it possible to have three separate phone numbers in one sheathed telephone wire.

(2) Green, red, black yellow, white-blue, blue-white wire looks like it should go under the heading of odd wire, but this is regular telephone wire often used in homes. (See (1) of four-conductor wire). The only difference in their use is that this wire will have the White and Blue wire as the third pair.

Eight Conductor

Eight conductor wire is the same as six conductor except for the fourth pair. (See (1) of six-conductor wire). The fourth pair is the white-brown, Brown-white pair. On the new electronic phone systems it is possible to get numerous lines (separate numbers)

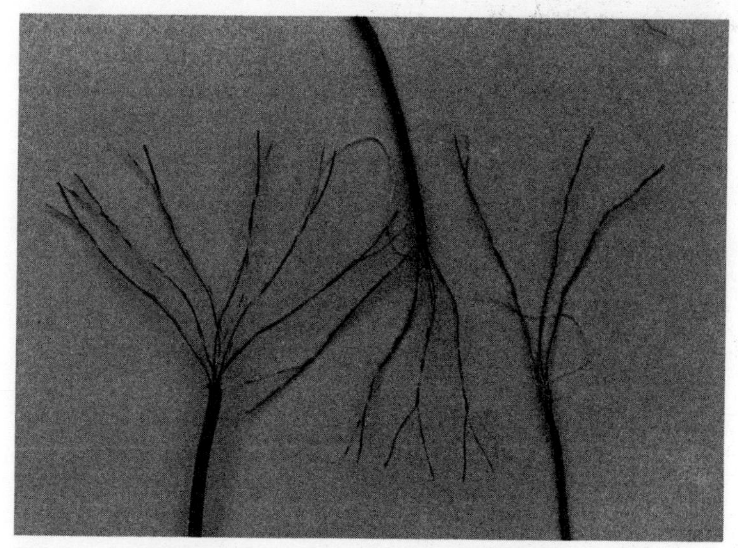

Fig. 7-4. All multiconductor wire except 25 pair or larger.

working on one electronic business telephone.

Teflon™ is one of the newest of the telephone wires. Its use is being stressed for safety reasons. One of the main causes of deaths in buildings is smoke inhalation from toxic fumes. Teflon™ wire does not give off heavy toxic fumes when it comes in contact with a flame. The color coding is the same as eight conductor wire. See Fig. 7-5.

The only drawback to Teflon™ wire is that is very difficult to strip down to the copper conductors. There is a special tool made to strip off the outer sheath but a sharp pocket knife will work on a limited amount of wire. For more information on the use of this wire or other wire or cable see the *National Electrical Code*.

Twelve Conductor

Twelve conductor (six pair) is like eight conductor but with two additional pairs. The white-slate, slate-white wires are the fifth pair and the red-blue, blue-red are the sixth. (See six- and eight-conductor wires for more information).

FIFTY CONDUCTOR OR TWENTY-FIVE PAIR

Twenty-five pair cable is used mostly in key phones with lights and hold capability. Cables of this size are put in business offices,

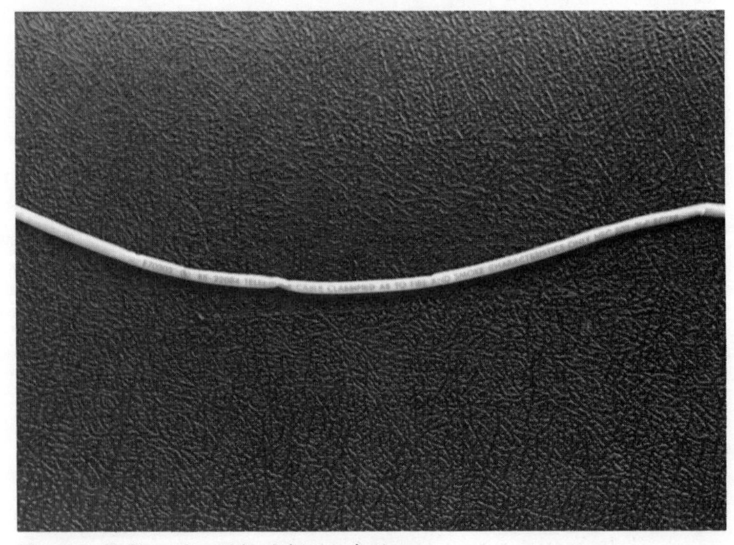

Fig. 7-5. Teflon wire with eight conductors.

but also occasionally, in houses for multiline use. The cable is terminated with a large female plug.

Six Pair

Fifty Conductor of Twenty-five Pair

White-blue blue-white	1st
White-orange orange-white	2nd
White-green green-white	3rd
White-brown brown-white	4th
White-slate slate-white	5th
Red-blue blue-red	6th
Red-orange orange-red	7th
Red-green green-red	8th
Red-brown brown-red	9th
Red-slate slate-red	10th
Black-blue blue-black	11th
Black-orange orange-black	12th
Black-green green-black	13th
Black-brown brown-black	14th
Black-slate slate-black	15th
Yellow-blue blue-yellow	16th
Yellow-orange orange-yellow	17th
Yellow-green green-yellow	18th
Yellow-brown brown-yellow	19th
Yellow-slate slate-yellow	20th
Violet-blue blue-violet	21th
Violet-orange orange-violet	22th
Violet-green green-violet	23rd
Violet-brown brown-violet	24th
Violet-slate slate-violet	25th

ODD WIRE

☐ Flat Cable
☐ Odd Color Wire
☐ Stranded

☐ Flat cable is not commonly used because it is expensive and its main value is hiding wire under carpet. It requires special mounting and attachments. It looks like computer flat cable. Sometimes, it is adhesive to stick to the floor to prevent slipping.

☐ With the divestiture of AT&T and changes in the telecommunications industry there will be people running wire in buildings with whatever can be found inexpensively. The wiring in a building could be anything which has wire conductors. Sometimes you will find places wired for telephones with wire that is not telephone wire at all. Sometimes the colors will not fit the color code for phone wire. The wire most commonly used instead of telephone wire is thermostat wire. Its conductors come in a variety of colors. You should use the most common color for telephones (the green and red wires) for dial tone. The phone company will probably not hook to these wires so they will go through an interface (See Chapter 2) to get the dial tone to you. Good Luck on this venture.

☐ STRANDED wire is never to be used for telephone station wire. (It is insulated wire with several fine conductors instead of the single #22 or #24 gauge conductor.) Telephone company people will not hook up to it. It is not an approved wire and can not be approved since it can not be terminated properly into telephone equipment. If you find it, rewire with the proper wire.

HOW TO STRIP WIRE

There are three main ways to strip wire from the outer sheath. There are two ways to strip quad or four conductor station wire and one way to strip multiconductor wire. The quad has nick and strip and nick and peel; the multiconductor has nylon string.

Nick and Strip

Tools Needed

☐ Diagonal cutting pliers

Steps

(1) This way of removing the outer sheath takes a little practice. At 8 inches from the end of the quad make a nick in the outer sheath. Do not nick the inside conductors.

(2) Grasp the quad or station wire behind the nick. Place the cutting edges of the diagonal cutting pliers on the nick.

(3) Grasp the outer sheath with the pliers and pull away from your body. The sheath will tear off and slide from the conductors. This will leave you with the four insulated conductors with no copper exposed (Fig. 7-6).

Nick and Peel I

Tools Needed

- ☐ Diagonal cutting pliers
- ☐ Longnose pliers

If you feel the first way of stripping is too difficult. Try it this way. It may be easier.

(1) Make a nick half way through the station wire at a point 4 inches from the end of the quad (Fig. 7-7).

(2) Bend the quad to expose the ends that were cut.

(3) Hold the 4 inches of end wire in one hand.

(4) With a set of long-nose pliers grab one of the wires that was cut and pull down the wire. This will peel the rest of the sheath off (Fig. 7-8).

(5) Trim the sheath to the length desired for the four insulated conductors.

Nick and Peel II

Tools Needed

- ☐ Wire stripper or diagonal pliers

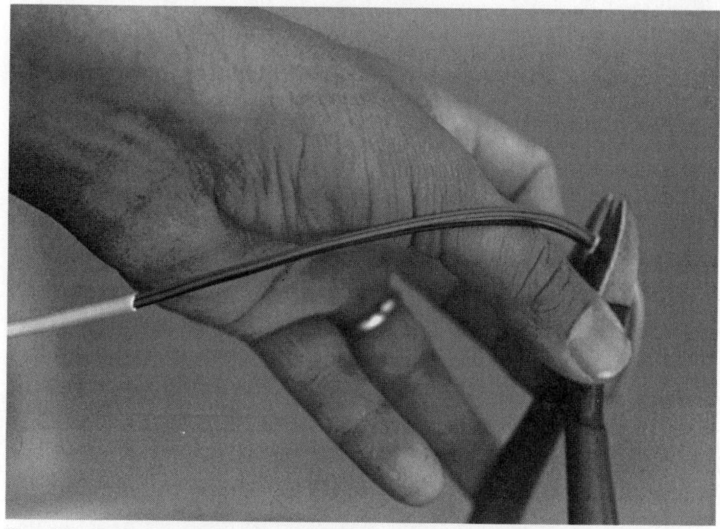

Fig. 7-6. Stripped station wire with diagonal cutting pliers.

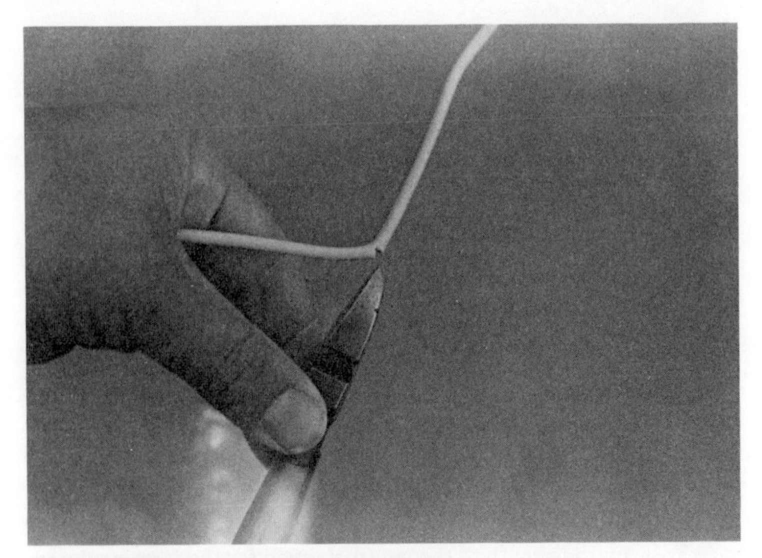

Fig. 7-7. Station wire nicked half way through.

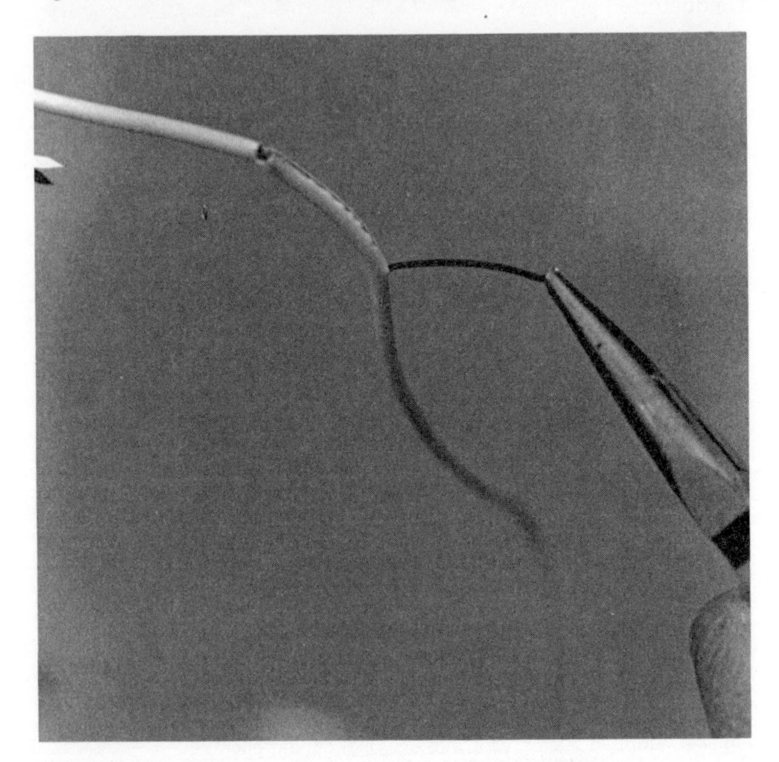

Fig. 7-8. Using long nose pliers to peel sheath off conductors.

This method is to be used on looped wire that you want to strip and attach to a jack uncut.

(1) Cut a small slit in the outer sheath without disturbing the inner wire. The slot should be just large enough to give access to the wires. See Fig. 7-9.

(2) Pull out one wire and use it to peel back sheath. See Fig. 7-10. See Chapter 8 for application.

Nylon String Peel I

Tools Needed

- ☐ Diagonal cutting pliers
- ☐ Longnose pliers

(1) Multiconductor wire has a nylon string to help peel off the outer sheath. With the diagonal cutting pliers, nick or cut the outer sheath and expose the inner insulated conductors.

(2) Locate the nylon string and peel the outer sheath to where you desire. Use a pair of longnose pliers to pull the string. Take care that the string does not get tangled with the conductors. If this does happen pull the conductors out of the way. See Fig. 7-11.

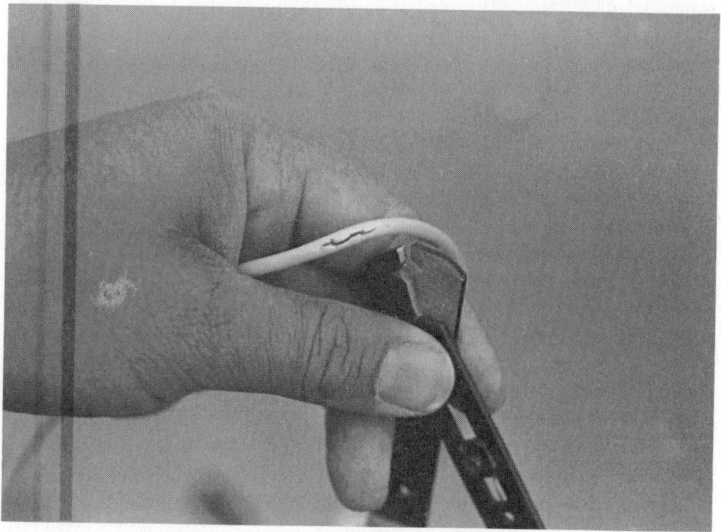

Fig. 7-9. Nicking looped wire without cutting conductors.

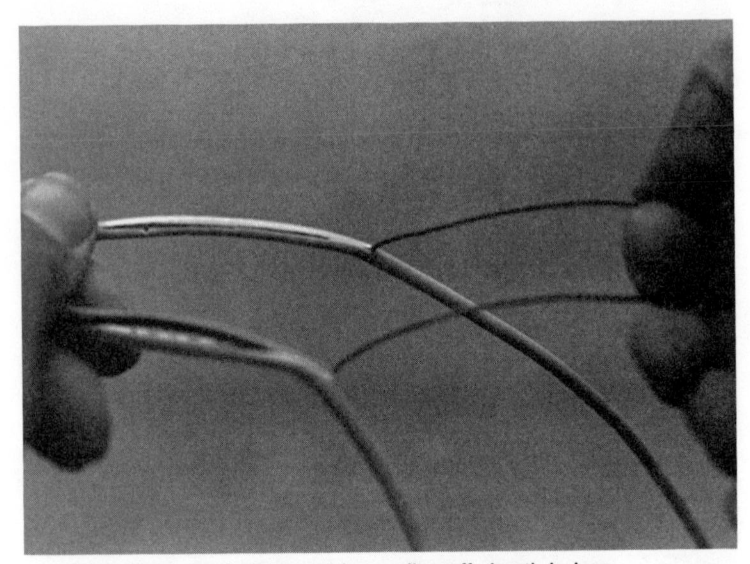

Fig. 7-10. Conductor being used for peeling off sheath in loop.

Nylon String Peel II

Tools Needed

- ☐ Something to nick wire.

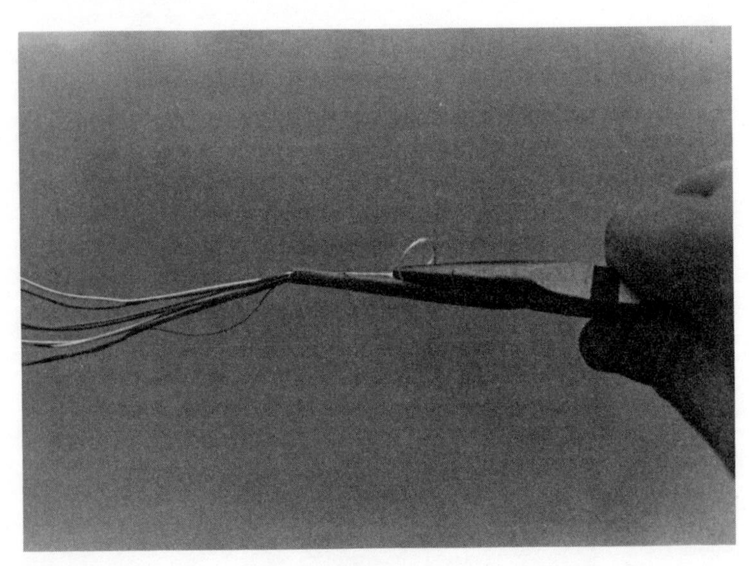

Fig. 7-11. String being used to strip off sheath with long-nose pliers.

This method is to be used on looped wire that you want to strip and attach to a jack, uncut.

(1) Nick the outer sheath just enough to find string.

(2) Pull string to open sheath so that you can get at wires. See Chapter 8 for application.

SPLICED WIRE

Note: Never splice outdoors. Only splice in areas not subject to disturbance and moisture.

There are three ways of splicing wire:

Bridged—into the connecting block of a jack or a wire junction. See Chapter 8

Directly spliced—into existing wire with wire connectors in a hidden place. See Chapter 10.

Pigtailed (worst type)—twisted together. If you have no choice, solder the wires and tape the ends with electrician's tape. See Chapter 11.

HOW WIRE IS RUN

Telephone wire can be run in different ways: (1) Looped or Multiple, (2) Home Run, or Direct Run (3) Combination. All that is needed for dial tone is the continuity of the wires for the circuit.

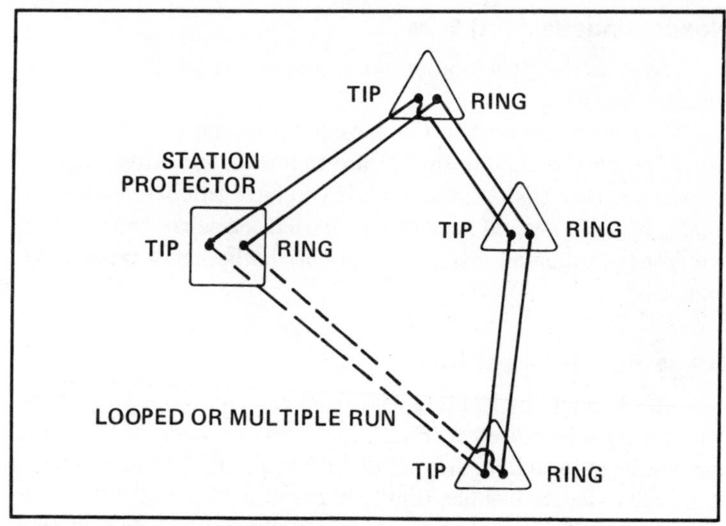

Fig. 7-12. Looped or multiple wire run.

Looped or Multiple Wire

This is the most common in prewired homes or apartments (Fig. 7-12). There are two paths looped wire may take:

☐ From the point of demarcation, past each consecutive outlet, to the furthest outlet, where it ends. Advantage: may be easier to run if the distance is great.

☐ From the point of demarcation, past each consecutive outlet, in a circle and terminating at the point of demarcation (back feeding). Advantage: If there is damage to the wire on the line, you can cut the wire in the middle to free part of the jacks, and reroute from the outer end. There are two ways to run the wire on either of these paths.

One Continuous Uncut Station Wire

☐ Starts at the point of demarcation and is run past each consecutive outlet through to the end.

☐ An uncut loop is pulled into each outlet.

☐ The wire is "made down" on the desired jacks and left looped for future use in the unused outlets. Advantage: You do not need to make down or splice each outlet to have continuity to jacks. Disadvantage: Limits the number of phone lines.

Noncontinuous (Cut) Wire

Starts at the demarcation point and is run to the first jack, where it is cut.

Runs from the first to the second outlet and is cut.

Repeat for second to third, third to fourth, etc. Advantage: It is easier to run. Disadvantage: (1) If you have trouble, it could be at any jack. You would have to isolate each one separately. (2) You would have to install a jack at each outlet whether you need one or not.

Home Run or Direct Run

This is made by running wire from each outlet or jack to the wire junction or protector (Fig. 7-13). This may occur if the home has not been prewired. Advantage: (1) With one line to one jack, it is easy to locate trouble. (2) If you prewire with multiconductor wire, and if the runs are made into a closet and terminated on a

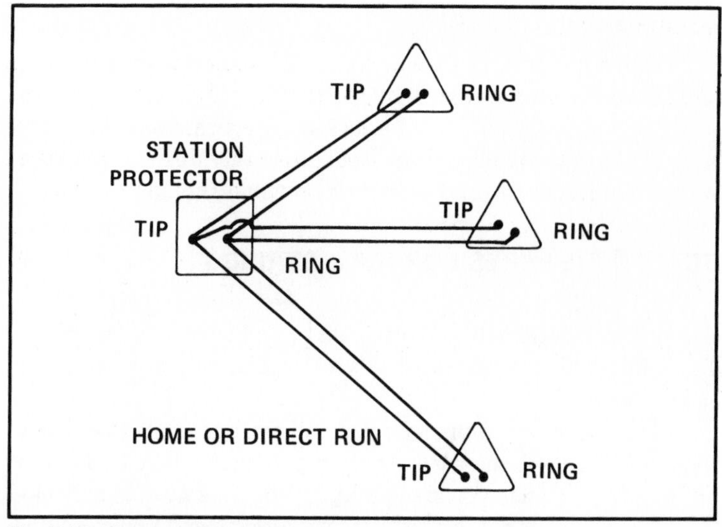

Fig. 7-13. Home or direct wire run.

termination block, you will later be able to install a multi-line telephone system. (You will need to tag and number each wire at both ends.) Disadvantage: a maximum of four wires can be terminated on any one protector or wire junction.

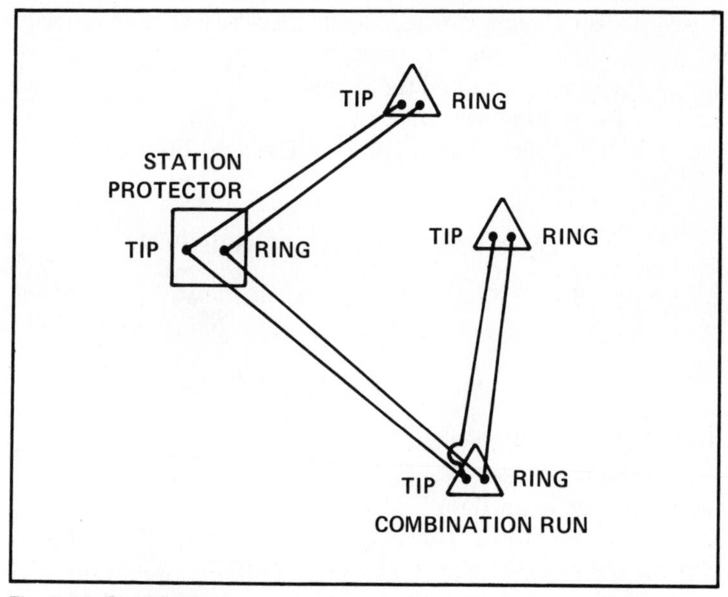

Fig. 7-14. Combination run.

Combination

This is just what it sounds like, a combined version of continuous and direct run (Fig. 7-14). Advantage: If a house is already wired and you want to install a jack in a separate area you may want to do this. Disadvantage: Trouble is *very* difficult to locate and very difficult to fix. I *do not* recommend this method.

HOW TO TIE WIRES FOR PULL STRINGS

There will be times when you may need to use an existing wire as a pull string. (A pull string is used to pull wire or cable in conduit but in this case an existing wire is used to pull new or secondary wire).

Note: Never force the wire if it will not pull. Pull it back and try again or retape the wire. Another person can help by feeding the wire to you when you are pulling it. You may want to pull wire in order to replace a damaged wire with new wire. You can "tie" the new one to the old one and pull the one one out. The new wire will then be in place. You may also want to pull wire in order to bridge onto a working jack. Example: You need to run the new wire through a difficult area in the wall from a working jack to the ceiling.

☐ Disconnect the jack.

☐ Tie on a length of wire to make the wire longer so you will not pull it entirely from the outlet.

☐ Pull the end up into the ceiling.

☐ "Tie" on the new wire.

☐ Pull the taped extension back through. Both wires will follow (Fig. 7-15).

Tools and Equipment Needed

☐ Diagonal pliers
☐ Electrician's vinyl tape

Methods of Tying

☐ Taping on the wire
☐ Twisted loops
☐ Double wrap loops

Taping Wires

To pull new wire or replace.

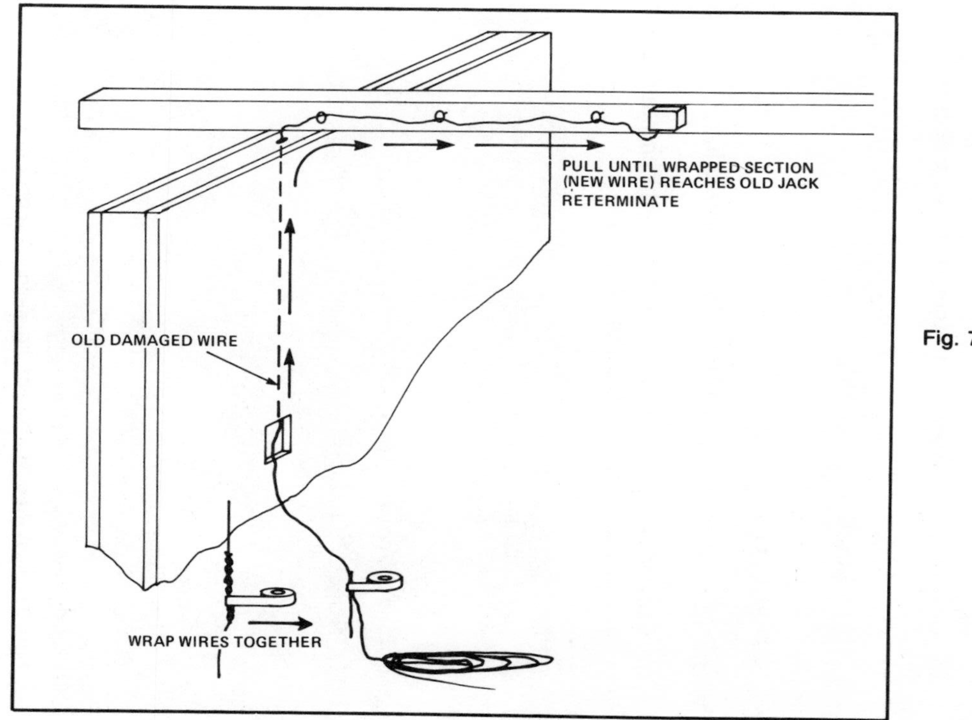

Fig. 7-15. How wire is used for a pull string.

(1) Cut both wires clean at the end (if you have plenty of wire).

(2) Overlap about six inches of one end of a spare wire, and 6 inches of the other end of the other wire.

(3) Wrap tape around wires as shown. After the first wrap keep the tape taut and stretched as you wrap.

(4) Tape beyond each end to be sure the wires do not snag on something when they are pulled (Fig. 7-16).

Twisted Loops

To replace wire if there is plenty of room to pull through.

(1) Cut end of wires clean.

(2) Loop wires through each other then twist ends around themselves.

(3) Tape wires from one end to the other being sure the tape is stretched as it is wrapped (Fig. 7-17).

Double Wrapped Loops

Use when there is limited space to pull through.

(1) Cut ends of both wires clean then strip the sheath of both wires not less than 4 inches from the end.

Fig. 7-16. Taping of new wire to existing station wire prior to pulling.

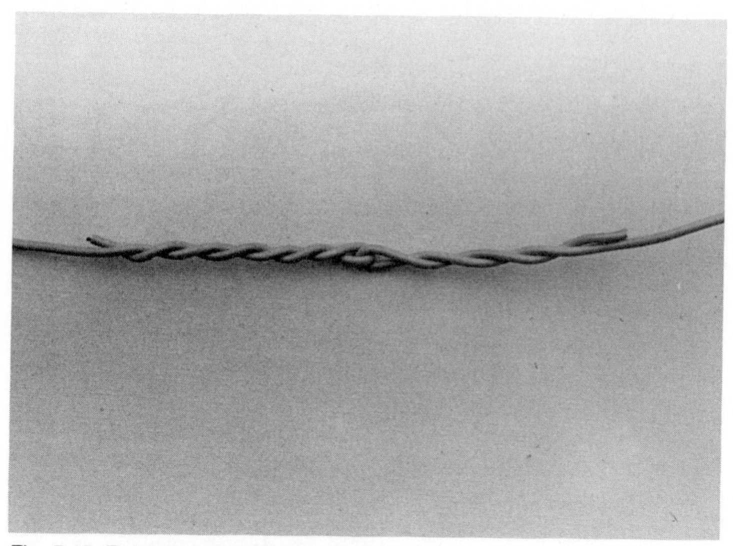

Fig. 7-17. Twisted loops without taping.

(2) Cut half of the conductors on each wire close to where the sheath is cut (Fig. 7-18).

(3) Take one wire and double loop it (as shown in Fig. 7-19 and 7-20) to the other wire. This will make a ring of about one and one half inches in diameter.

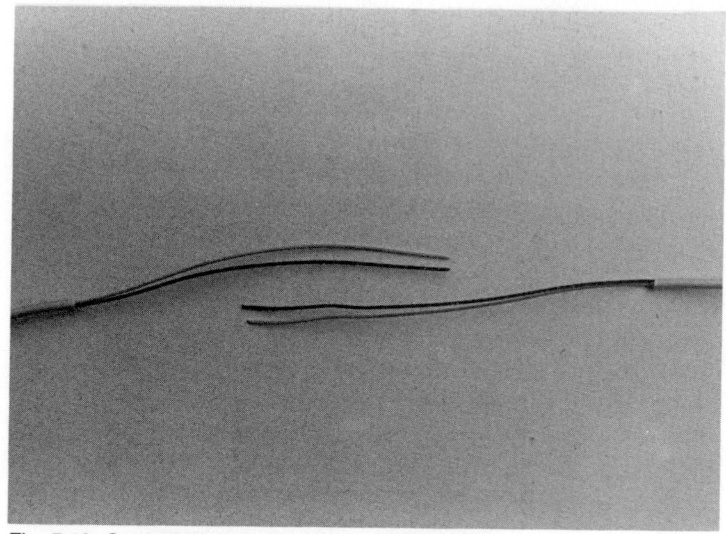

Fig. 7-18. Cut half of the conductors at the sheath.

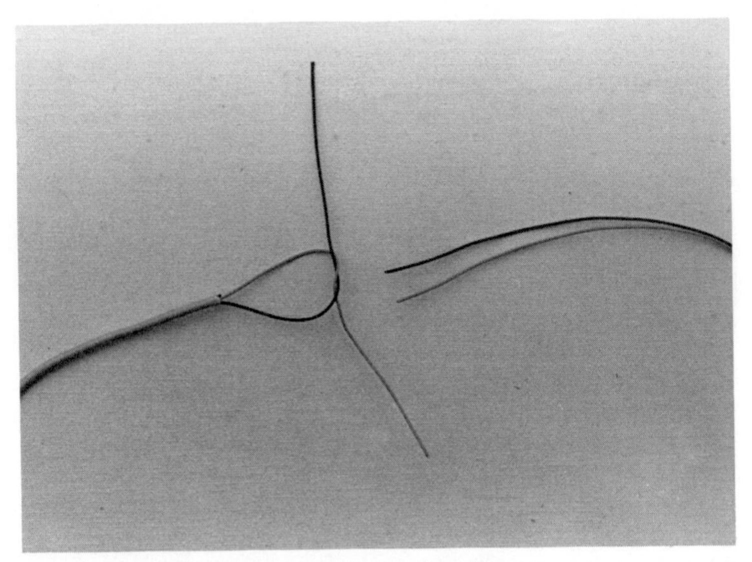

Fig. 7-19. Twist set of wires to begin to weave loop.

(4) Do the same with the other wire looped through the first ring (Fig. 7-21).

(5) Flatten the rings until they are no wider than the sheath (Fig. 7-22).

(6) Tape the wires together being sure the tape stretches as

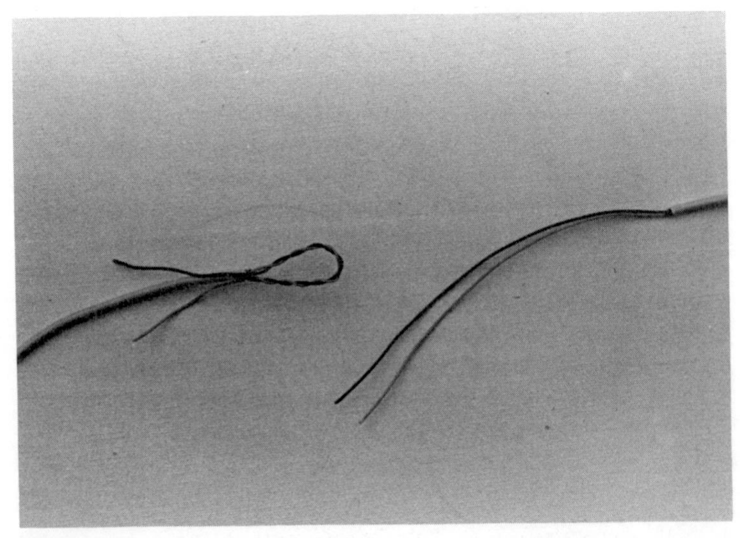

Fig. 7-20. Finished loop.

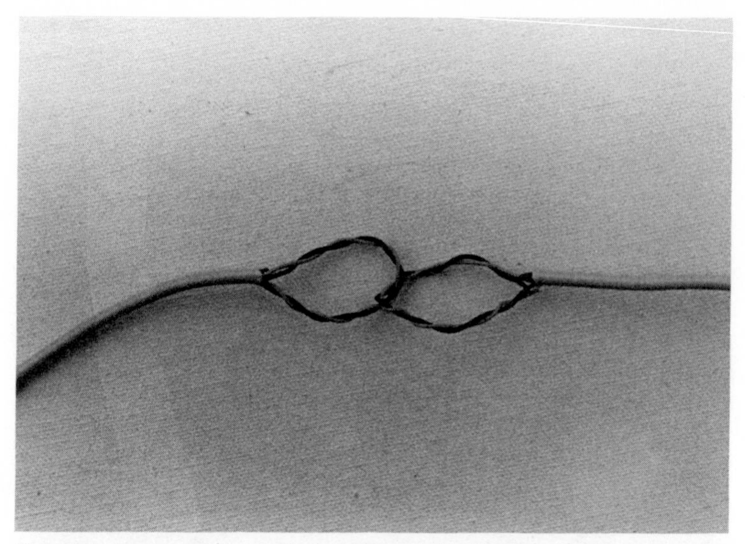

Fig. 7-21. Loop second set of wires through the first loop and weave in the same manner.

it is wrapped. Go over the length of both wires where the sheath is cut. Put only one wrap of tape over the length to give a small diameter with which to work (Fig. 7-23).

DO'S AND DON'TS OF WIRE RUNNING

These are some special tips to help you run wire safely, effectively, and attractively.

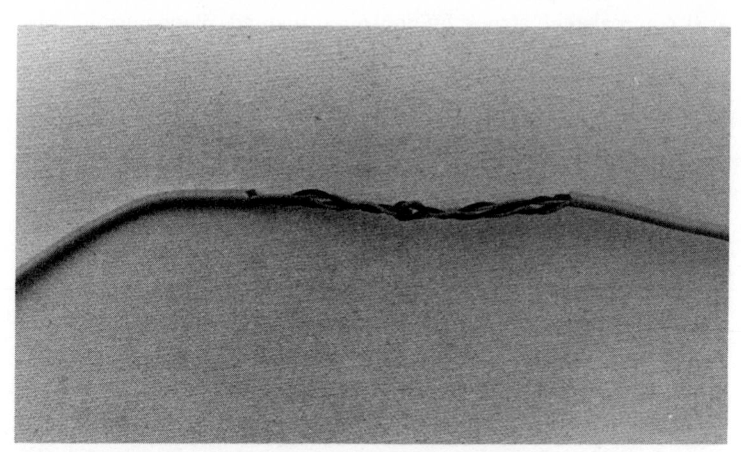

Fig. 7-22. Flatten rings to same diameter as station wire.

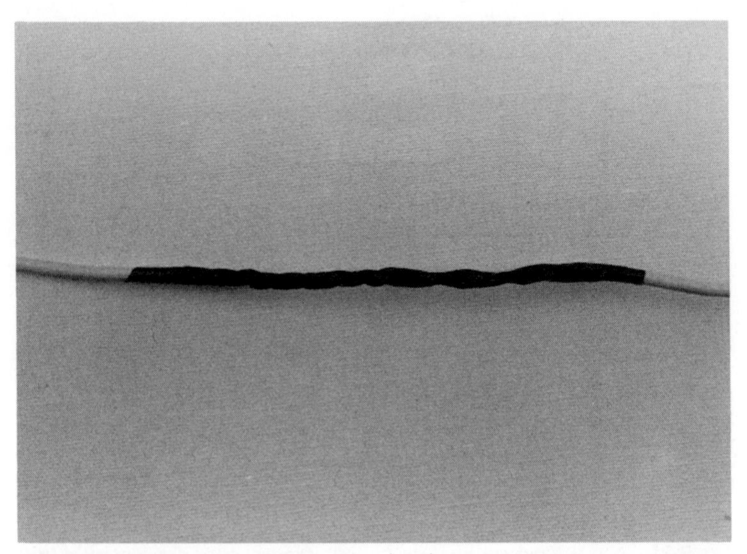

Fig. 7-23. Tape across the length of wire from one sheath to the other sheath.

Don't

Safety

(1) Never drill without knowing what is in front of the drill. You could drill into power lines, gas or plumbing.

(2) Do not run telephone wire in the same conduit, outlet boxes, or compartment with electrical wiring.

(3) Do not span or run wire in the air. Someone may run into it. It could also be struck by lightning. Instead, place wire in conduit between buildings. Be sure it is attached to the telephone service protector.

(4) Do not install telephones or jacks in areas that are damp or wet. The user could get shocked. Note: Despite this caution, people still want a phone near areas that have these hazards: Never use a phone when you are damp or when a storm is in progress.

(5) Do not run telephone wiring next to bar electrical wiring or any form of high-voltage transmitting signaling devices.

Trouble

(1) Keep wires from being abraded by double wrapping electrical tape around the area where damage is likely.

(2) Do not hang anything from telephone wire (lighting fix-

tures or other decorating items).

(3) Install wires where there will not be strain, pinching, or cutting (such as window sills or door jams.)

(4) The wires should never be pig-tailed or just twisted directly together. (Use connectors, jacks, or wire junctions—see Chapter 1.)

(5) Do not splice telephone wire outside of a home or building. Since the sheath is damaged, moisture will eventually get in and cause trouble. See Chapter 11.

(6) Do not run wire over the roof of a building.

(7) Do not fasten wire to any fixture that is not permanent.

(8) Do not run wire through another party's property (apartments or condos).

(9) Never run wire in front of a sign, or other fixture that is movable.

(10) Never run wire in a hot or steamy location.

(11) Never run wire under carpet except at corners or with flat telephone wire for use under carpet.

(12) Do not attempt to run wire outside when the temperature is very cold. Wire shatters easily at 20 degrees F. If you live in a climate that is very cold, run wire inside or in conduit.

Do

(1) Do run wire the shortest or most direct path.

(2) Do run wire horizontally or vertically in a straight line. This also means right angles.

(3) Do run wire (if on the outside of a home) on the sides or the back.

(4) Do use wooden surfaces for attachments.

(5) Do use telephone wire (not just any old wire) for runs.

(6) Do leave some slack in wire to allow bridging of wire in future.

(7) Do limit wire runs to 250 feet if it is four conductor and is carrying two service lines.

(8) Do hide wire when possible.

(9) Do terminate wire to phone jacks properly.

(10) Do have all proper tools and equipment before starting job.

(11) Pull bulk wire from the middle of the coil as shown in Fig. 7-24.

LINES OF ILLUSION

You will be most satisfied if, in addition to personal safety and

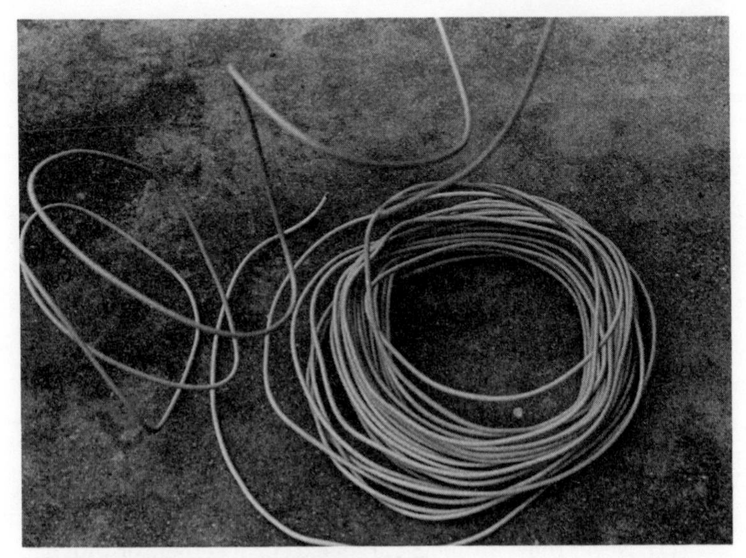

Fig. 7-24. Pulling bulk wire from the middle.

a well functioning telephone, the wire that you run is not an eyesore. Sometimes you will be able to run the wire inside the walls, the attic, or the basement. When this is not feasible, you should try to fool the eye. There are several tricks for camouflaging wire.

(1) Do not run wire at eye level. Run it high or run it low.
(2) Do not run it across the front of a building.
(3) Do not run it across a bare, flat wall.
(4) Do not run wire diagonally unless there is a feature of that shape that you are following.
(5) Hide the wire under eaves.
(6) Run wire vertically up corners.
(7) Run it along side window frames, door sills, etc.

In other words, follow the natural features of the building. Choose areas that are not in the usual line of vision (high, low, back, etc.).

Chapter 8

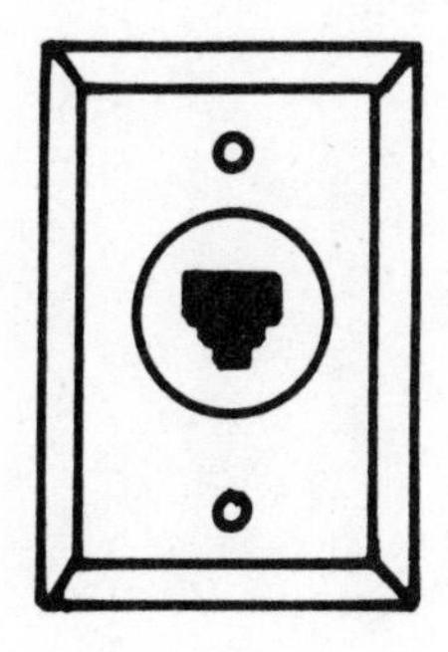

Jacks

A JACK IS THE DEVICE USED TO CONNECT THE STATION WIRES to your telephone. It usually consists of two parts (even if contained in one unit); the connecting block, or base, and the cover. The connecting block mounts to the wall. It contains the terminals (usually screws, but sometimes split beam or other) to which the station wires are to be connected. The cover is already connected to the terminals and also contains an outlet into which the phone plug snaps.

For many years, there was only one type of jack, the hardwire. See Fig. 8-1. It did not have a disconnect outlet. Instead, the telephone was terminated into the jack permanently, just as the station wires were. The invention of the four prong jack made it possible to have several jacks and only one phone which could be moved wherever it was needed. This applied only to the desk phone but not to wall phones, into which the station wire was wired directly without a jack. The outlet of the four prong jack had four female receptacles and the phone plug has four male prongs. See Fig. 8-2. This arrangement was very durable. It still worked after many pluggings and unpluggings. For this reason, some people still prefer and use the four prong jack.

More recently still, a new variation on the movable jack was invented. This new jack, called the modular jack, has an outlet that fits with a squeeze tab plug. See Fig. 8-3. Although it is not quite as durable as the four prong, it is very quick and easy to plug and

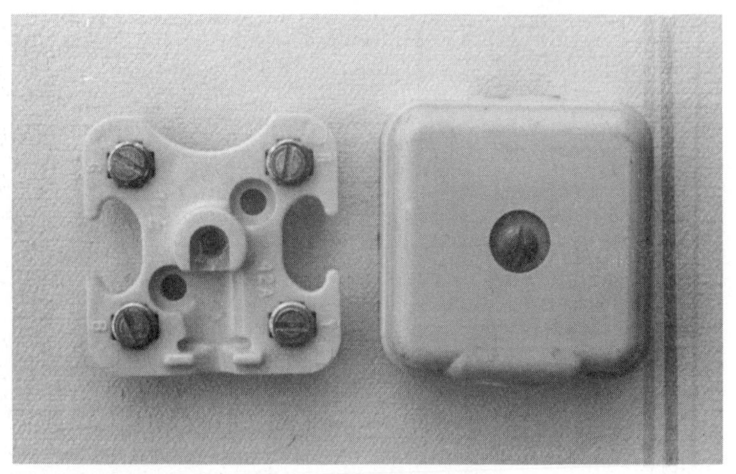

Fig. 8-1. Hardwire 42A connecting block with cover.

unplug. The phone itself may not even be a single unit. The cords and receiver may snap on and off with the modular connection and thus can be replaced. The wall phone has also become mobile. The phone slips down over and makes contact with a jack that is attached to the wall. The important thing about the modular connections is that they are now standardized. Because of the divestiture of AT&T from Bell System, and the changes made by the FCC about phone ownership, the modular jack has become the standard connection for most telecommunications. It has made it possible

Fig. 8-2. Example of four prong jack and plug.

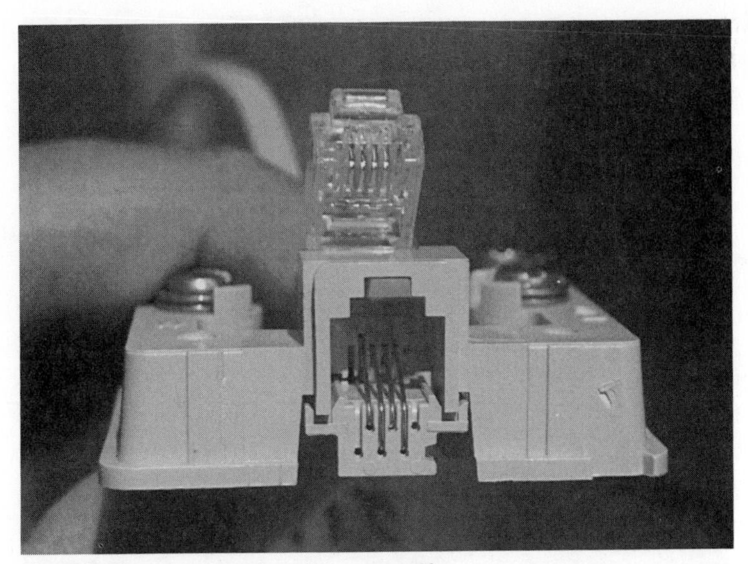

Fig. 8-3. Close-up view of modular connection.

for you to use a large variety of makes and models of phones (AT&T's and anyone else's that is registered with the FCC).

There are so many different models of jacks that I could not begin to show them all here. Even if I could, they are changing so rapidly that the list would be outdated before you got to the store. Instead, I will tell you what some of the variables are, so that you can examine the jacks available in the store and determine which type you want. There are basically three types of jack: (1) a surface or baseboard jack (which attaches to the top of the baseboard), (2) the flush or outlet jack (which fits into the wall like an electric outlet), and (3) the wallphone jack (over which a wall phone slips). There are some differences in the methods of mounting these to the wall. There are essentially two ways to terminate wire: either with screws or with split beams. Finally, the covers are somewhat different. All of this is covered in some detail in the Installation section of this chapter.

USOC CODES

USOC codes are used to identify certain qualities in jacks.

RJ11C—is the USOC for a standard desk modular jack. It can be a flush or surface type. Into it, can be plugged any form of telephone equipment that only needs a single telephone line. This is the most common type for a residence.

Table 8-1. Telephone Jacks.

USOC	Jack Description	Installation	Use
RJ11C	modular surface or baseboard type-see Fig. 8-8 or 8-17	On baseboard, moulding, any area without an outlet box, on solid wall surface, or on a stud where wall covering is not sturdy.	One telephone line or number in operation per jack-Desk type
	modular flush type-see Fig. 8-22	In an outlet box and possibly directly on a wall if hollow and hole made. (some types on surface)	
RJ11W	modular wall phone mount type-see Fig. 8-25	In outlet box and directly on wall whether hollow or solid.	One telephone line or number in operation per jack-Wall phone type
RJ14C	modular surface or baseboard type-see Fig. 8-8 or Fig. 8-17	On baseboard, moulding, any area without an outlet box, on solid wall surface, or on a stud where wall covering is not sturdy.	Two telephone lines or numbers in operation per jack-Desk type

	modular flush type-see Fig. 8-22	In an outlet box and possibly directly on a wall if hollow and hole made. (some types on surface)	
RJ14W	modular wall phone mount type-see Fig. 8-25	In outlet box and directly on wall whether hollow or solid	Two telephone lines or numbers in operation per jack-Wall phone type
RJA1X	modular adapter-see Fig. 8-4	Plugs into four prong or pin jack to convert to modular jack	Will make modular, can be considered RJ11C
RJA2X	double modular adapter	Plugs into existing modular jack	Allows one modular jack to become two in the same area.

RJ11W—is the same as the RJ11C except that it is a wall mount type instead of a desk.

RJ14C—is like a RJ11C except it provides two telephone lines. This jack can be used for a business or a residence (using a special phone or a regular phone with an adapter) to receive two different numbers on a single telephone.

RJ14W—is the same as a RJ14C except it is for a wall phone.

RJ25C—is used mostly on business systems that require three lines working from one jack. These come with six contact rails instead of the four in the standard modular jack. There is no difference in the size of the plug except for the pair of extra contact rails.

RJ31X—connects special telephone equipment ahead of all other telephone equipment. It cuts off all telephone equipment in front of it. It is most commonly used for automatic dialers in burglar alarms or answering machines.

CHART OF JACKS

If you are installing jacks where no service exists, you will probably want to install modular jacks. You may, however, be working on an already existing system that has not yet been changed to modular equipment. Some equipment can be adapted, some should be converted, and some must be completely replaced or be installed as a new extension. See Table 8-1.

ADAPTATION

Adaptation from four prong to modular or from modular to four prong is as simple as plugging one plug into another. The adapter has either a four prong plug and a modular outlet, or a modular plug with a four hole outlet (Fig. 8-4).

The USOC numbers for two adapters are: RJA1X—converts a four-prong or four-pin jack into a modular jack. It makes a four-prong jack into a RJ11C. This adapter allows any standard modular cord to plug to the four prong. RJA2X (Duplex Jack)—converts a standard modular jack or RJ11C into a jack that you can plug two separate modular cords. This allows you to have two phones or a phone and an answering machine in the same place.

CONVERSION

Conversion is a term that is used in the telephone trade to mean

Fig. 8-4. Adapter for four prong to modular.

any form of installing a new jack in place of an old one including the removal of an old one down to the station wires. Since, however, this book breaks the steps down into installing wire and installing jacks, we will use the term conversion to mean taking a new modular cover (containing a modular jack) and terminating it to an old existing 42A connecting block.

Converting a hardwire baseboard jack to a spadetipped modular jack or to a snap or button modular jack. See Fig. 8-5.

Tools

☐ Screwdriver

Equipment

☐ Modular jack cover (check to be certain it fits a 42A connecting block).

Safety

If you have dial tone, take the handset off of the hook on at least one phone. Only handle one wire at a time.

Steps

(1) Unscrew the center screw of the cover. Remove the cover and discard it.

(2) Disconnect the old wires. (If you are converting to a snap

or button modular jack, you can skip this section. Just cut them off where the wire connects to the spadetip, and fold them out of the way.)

☐ Look at the spade tipped ends (of the telephone mounting cord) terminated under the screws. You will either see four wires (green, red, black, yellow) or three wires (green, red, yellow). The block will probably have the letters G, R, B, and Y beside the screws on which those colored wires are terminated. If not, you need to mark them.

☐ Loosen the screws.

☐ Remove the spade tips and wires of the old mounting cord.

(3) Connect the new wires.

☐ Match the color designated screws with the correct colors of the new wire. If the colors are not G, R, B, Y see Chapter 7.

☐ Place the new spadetips so that they point downward

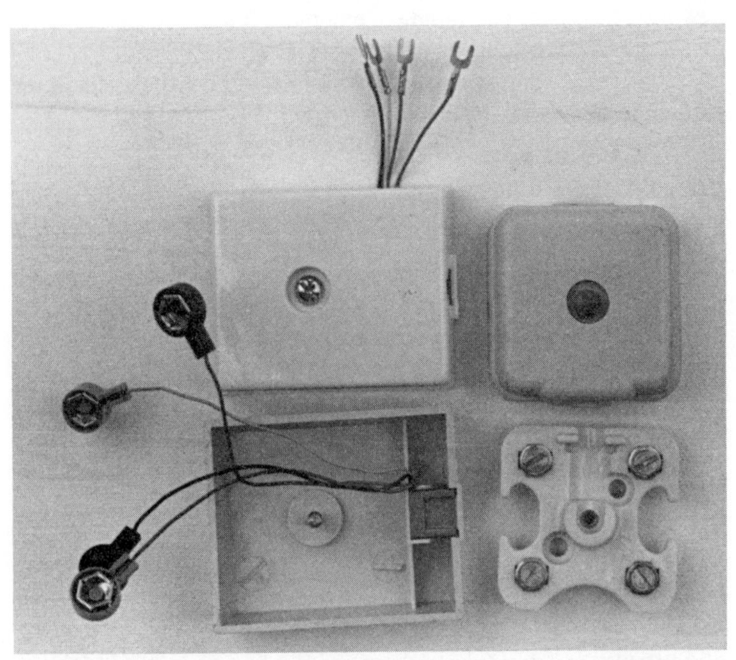

Fig. 8-5. Spade tip and snap or button modular baseboard jack covers.

and upward at one another as shown in Fig. 8-6. If you are converting to a snap or button modular jack, snap the buttons on and skip the next step.

☐ Tighten the screws.

(4) Check the phone.

☐ Plug in the modular phone.

☐ Replace receiver(s) on hook(s).

☐ Wait a moment, then lift the receiver and listen for dial tone. If you do not hear dial tone, check the wiring for looseness or breaks. Once you have dial tone, try to call out. If the phone is push button or Touch Tone™ and if you cannot dial out, reverse the two wires (at the jack) that supply dial tone (usually red and green).

Try again.

(5) Put on the new cover.

☐ Unplug the mounting or base cord.

☐ If the wires are long enough: Turn the cover three or four times so that the wires between the cover and the block are twisted. Wrap the wires behind the block as shown in Fig. 8-7. This will keep the wires from being pinched or

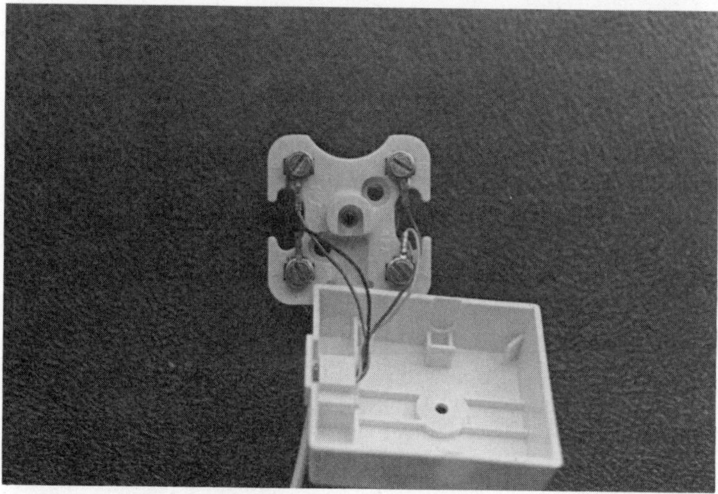

Fig. 8-6. Position of spade tips on screw terminals.

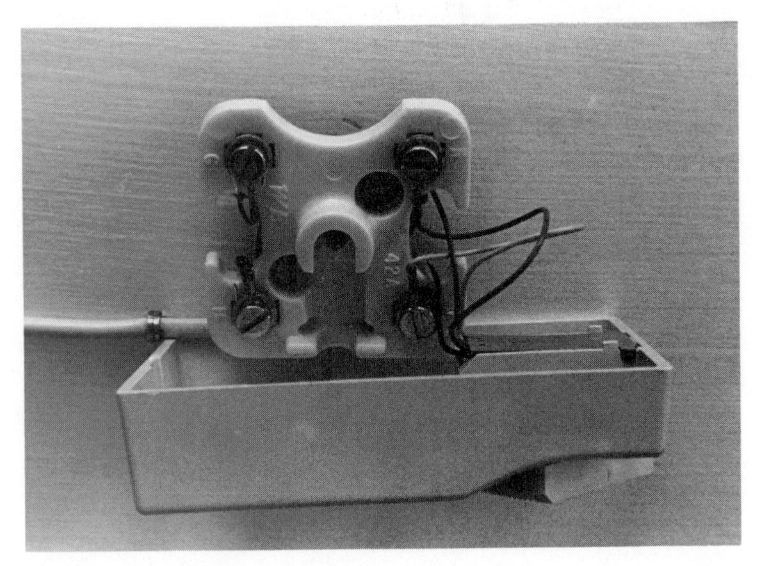

Fig. 8-7. Wrapping cover wires.

cut when the cover screw is installed.

☐ Place the cover so that the modular connection does not point upward (to prevent the collection of debris). Be certain that there is room to plug and unplug the cord.

☐ Do not force the screw in. These covers tend to go in a little crooked and if you are not careful you might strip the threads.

INSTALLATION

In the context of this book, installation means connecting a jack to the station wire preparatory to plugging in the telephone. It is assumed that when you begin this section you have already decided on the location, purchased the jacks, and run the wire.The four stages to the installation of a jack are, positioning, mounting, terminating, and covering.

Tools

☐ Screwdriver

☐ Longnose pliers

☐ Diagonal cutting pliers

☐ 1/4-inch nut driver, power drill (optional), or push drill (optional)

Equipment

☐ One of the following jacks:

(a) 42A connecting block and cover
(b) One unit jack
(c) One piece unit except cover is separate
(d) Spring cover with own base.

☐ 3/4-inch or 1-inch hex head screws, 3/4- or 1-inch pan head screws (optional)

Surface: (1) 42A connecting block and cover (2) One unit jack (3) One piece unit except cover is separate (4) Spring cover with own base.

42A Connecting Block and Cover (Surface) (Fig. 8-8).

Positioning:

Position the connecting block so that the green terminal is in the upper left corner and the yellow is in the lower right (Fig. 8-9).

Mounting

(1) If the surface to which you plan to mount the jack is flimsy, locate a stud.

Fig. 8-8. Modular 42A connecting block and cover.

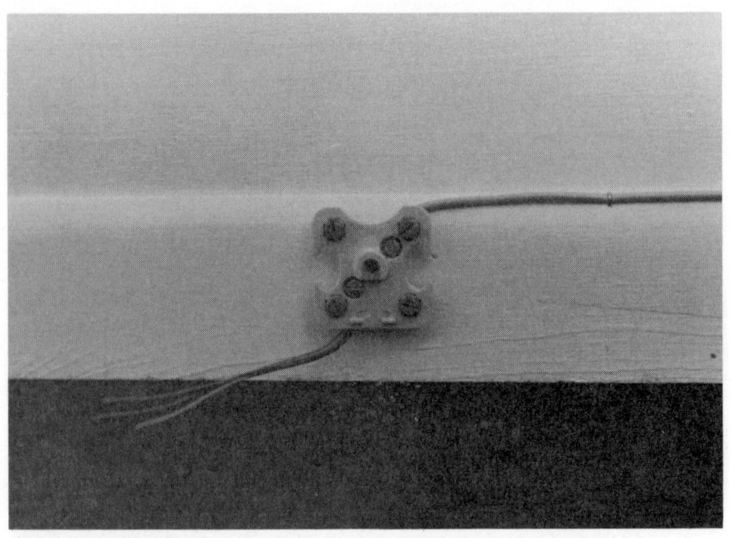

Fig. 8-9. Proper mounting for 42A connecting block on baseboard.

(2) To mount a block

☐ If the wood is not likely to crack:

(1) Attach a hex head screw (3/4-inch or 1-inch) with a 1/4-inch nut driver.

(2) Finish with an electrician's screwdriver.

☐ If the wood is brittle:

(1) Drill the holes with a 3/32-drill bit.

(2) Attach with 3/4-inch or 1-inch pan head screws.

Termination

(1) Remove 8 inches of sheath.

(2) Bring the wire into the channel, the beginning of the sheath should be in the middle of the back of the jack (Fig. 8-10).

(3) Staple over the wire (carefully!) right against the block.

(4) Wrap the black and red pair of wire around the back of the block one and one half times clockwise. The red should end up on the top right, and the black on the bottom left.

(5) Wrap the green and yellow pair of wires around the back of the block one and one half times counterclockwise. The green

should be on the top left and the yellow on the bottom right. See Fig. 8-11.

(6) Trim the wire to one inch.

(7) Strip the insulation close to the termination screws.

(8) Loosen the screws two and one-half turns.

(9) Connect the wires between the washers, with a clockwise hairpin turn, to the screws.

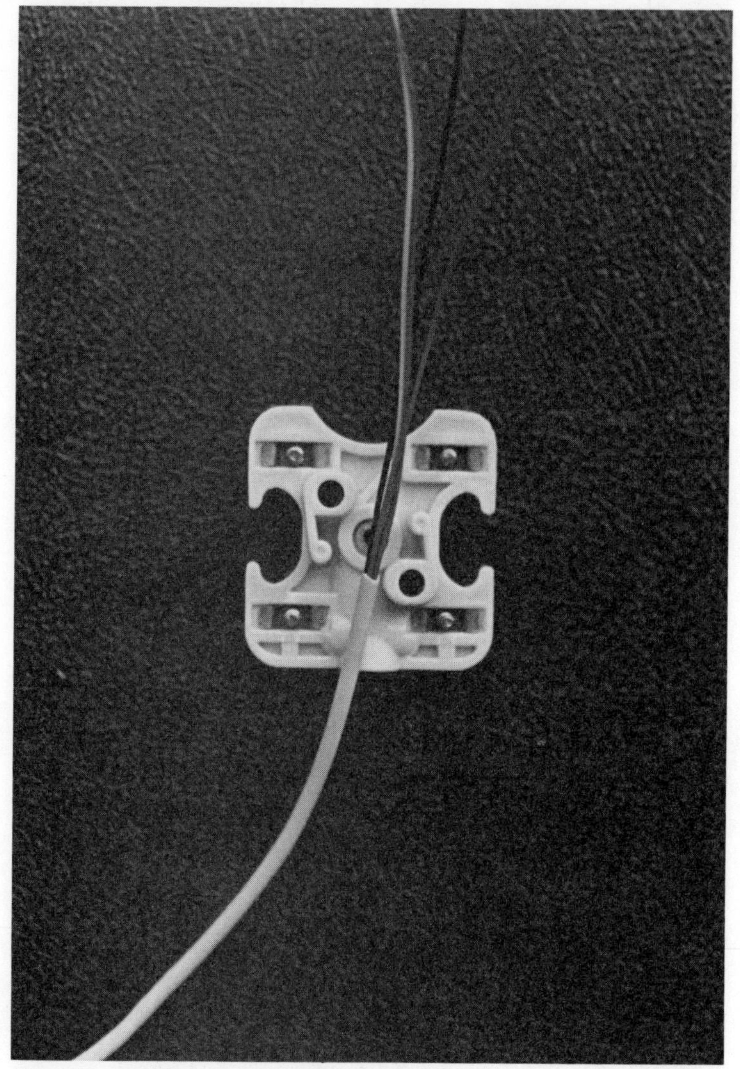

Fig. 8-10. Placing wire in channels on back.

Fig. 8-11. Wrapping of wires behind connecting block and into proper terminals.

(10) Trim off the excess wire (leave no more than 1/8 inch). See Fig. 8-12.

Installing Cover

(1) Wrap the wires, if they are long enough, behind the jack. Refer to Fig. 8-7.

(2) Be careful, as you screw in, not to pinch or cut the wires. See Fig. 8-13.

To Use a 42A Surface Jack as a Splice

(1) Remove cover (do not tighten screws).

(2) Move each old wire from between the washers to beneath the bottom washer next to the connecting block.

(3) Remove 8 inches of sheath from the new wire.

(4) Group the Black and Red wires together, and the Green and Yellow together. See Fig. 8-14.

(5) Wrap B and R clockwise and G and Y counterclockwise, one and one-half times to the corresponding color terminals.

(6) Trim excess wire to one inch beyond the screw.

(7) Strip one inch of insulation.

(8) Place the new wire where the old was (between the washers) with a hairpin clockwise turn.

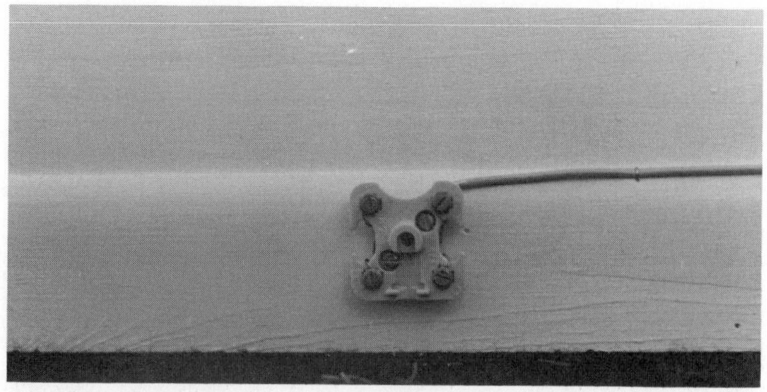

Fig. 8-12. Wire left trimmed and terminated on connecting block.

Fig. 8-13. Cover screwed on 42A connecting block.

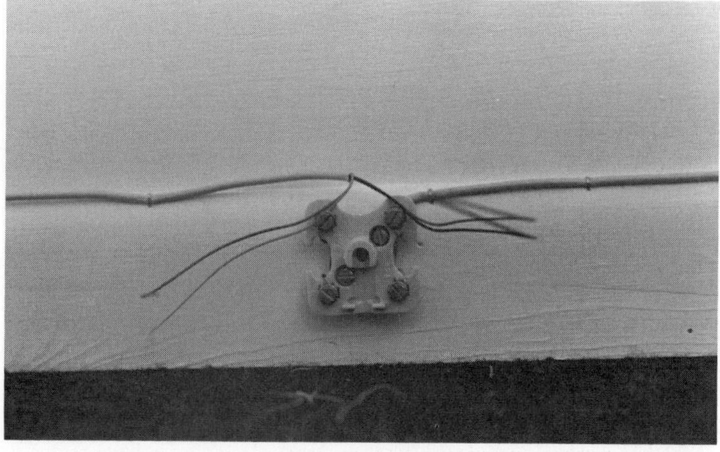

Fig. 8-14. Station wire with sheath stripped and pairs separated for wrapping.

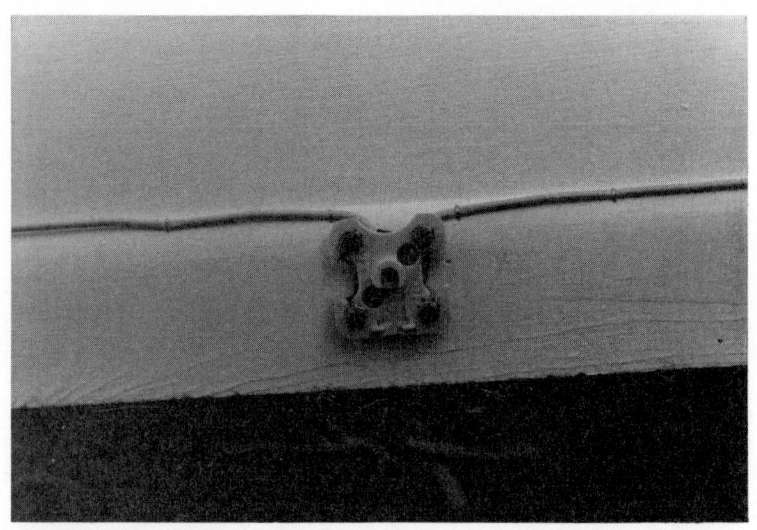

Fig. 8-15. Wires wrapped around connecting block to be spliced, ready for cover.

(9) Trim excess wire. See Fig. 8-15.
(10) Replace the cover.
(11) Snug down the screw. See Fig. 8-16.

ONE UNIT JACK (Fig. 8-17) (Surface)

Positioning

(1) Make certain that there is room to plug and unplug the phone connection.

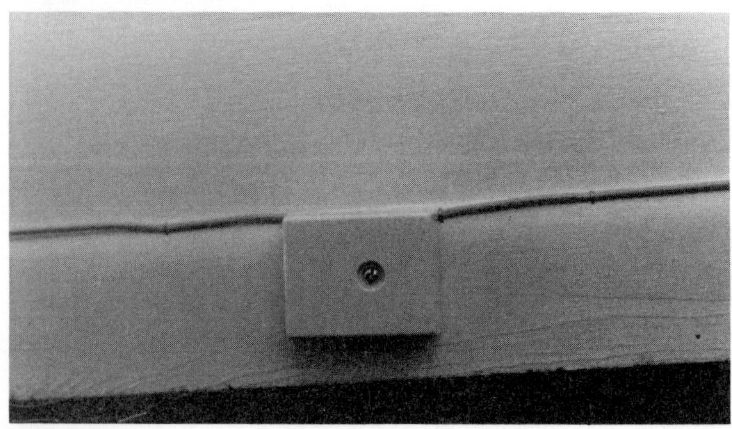

Fig. 8-16. Cover is screwed down on 42A connecting block.

Fig. 8-17. One unit baseboard jack.

(2) Turn the contact rails so that they do not collect debris.

Termination

This step is before mounting for this type of jack:

(1) Peel two inches of sheath from the wire.
(2) Remove one half inch of insulation from each wire.
(3) Loosen screws two turns.
(4) Terminate each wire, with a clockwise hairpin turn, according to the color marked on the jack.
(5) Snug down the screws.
(6) Trim excess wire.
(7) Stuff extra wire in the space behind the jack.

Mounting

(1) Pry out breakout tab to allow wire to pass through.

(2) Leave 1/2 inch to 1 inch of sheath in the jack.
(3) Screw the block on with the provided screws.

There is no cover to attach.

Block and Connection are in One Piece

Cover is separate and is only a cover (no wires) (Surface) Fig. 8-18.

Positioning

(1) Pry the cover off (do not damage the contact rails and the outlet).
(2) Make certain that there will be room to plug and unplug the phone cord.
(3) Turn the contact rails so that they do not accumulate dust and debris.

Mounting

(1) If the wood is not likely to crack:

- ☐ Attach a hex head screw (3/4-inch or 1-inch) with a 1/4-inch nut driver.

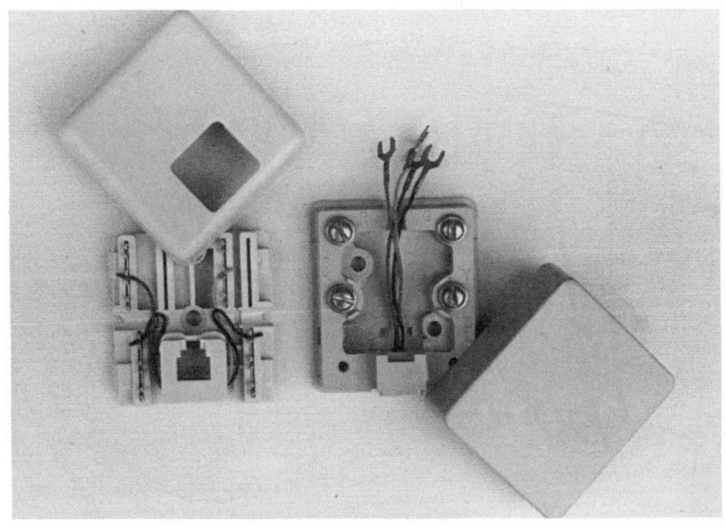

Fig. 8-18. One piece plus cover baseboard jacks.

☐ Finish with an electrician's screwdriver.

(2) If the wood is brittle:

☐ Drill the holes with a 3/32-drill bit.
☐ Attach with 3/4-inch or 1-inch pan head screws.

Termination

Screw Type

(1) Peel 2 inches of sheath from the wire.
(2) Remove 1-half inch of insulation from each wire.
(3) Terminate each wire, in a clockwise hairpin turn, according to the color marked on the jack.
(4) Snug the screws.
(5) Trim excess wire.
(6) Leave 1/2 to 1 inch of sheath in the jack.

Split Beam Type

(1) Remove 2 inches of sheath.
(2) Do not remove insulation.
(3) Match the wires to the corresponding colors.
(4) Push the wire into the terminals snugly with longnose pliers.

Cover

(1) Break out tab (if applicable) with longnosed pliers, to allow wire to pass.
(2) Stuff excess wire in jack.
(3) Snap on the cover.

Spring Cover Jack

Humidity and dust proof with its own base (surface) Fig. 8-19.

Positioning

(1) Make certain there is room to plug and unplug the phone cord.
(2) Place horizontally with the terminals on the left.

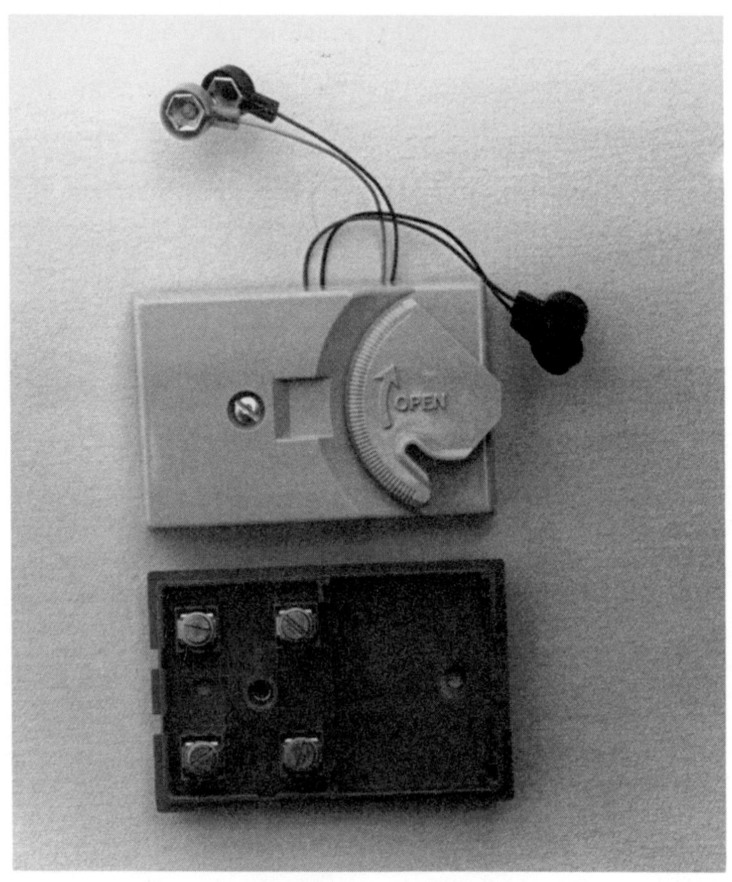

Fig. 8-19. Spring cover type baseboard jack (protective cover).

Mounting

(1) If the wood is not likely to crack:

☐ Attach a hex head screw (3/4-inch or 1-inch) with a 1/4-inch nut driver.
☐ Finish with an electrician's screwdriver.

(2) If the wood is brittle:

☐ Drill the holes with a 3/32-drill bit.
☐ Attach with 3/4-inch or 1-inch pan head screws.

Termination

(1) With a pair of long nosed pliers, pry out the break out tab from the side where you bring in the wire.

(2) Strip three inches of sheath from the wire, leaving one inch of sheath inside the connecting block.

Screw Type

(3) Remove 1/2-inch insulation from the ends of all the conductors and wrap them in a clockwise hairpin turn around the screw terminations. Do not let the wires go between the threads or the washers.

(4) Place the spadetips under the screw termination (between the screw head and the first washer). Tighten down gently. See Fig. 8-20.

Split Beam Type

(3) Do not strip.
(4) Insert the wire into the ports.
(5) Tighten screws.

Cover

Screw the cover on with the modular connector on the right.

Split Beam

Fig. 8-20. Terminating spadetips.

Snap the buttons on to the corresponding colors. Screw the cover down with the modular connection on the right.

Splicing with this type of jack is the same as installing except you will use all of the breakout tabs and all of the ports (corresponding colors).

FLUSH (OUTLET BOX OR NO OUTLET BOX)

Tools

- ☐ Screwdriver
- ☐ Longnose pliers
- ☐ Diagonal cutting pliers

Equipment

Jack (If you do not have an outlet box, choose a flush jack with a bracket that will lie flat against the wall.) If you have a blue plastic outlet box, see Note 1 in Chapter 3.

Wall anchors (two).

Positioning

(1) If you are using a bracket, consult the directions on the package (or the way the block and bracket are connected in the package). If the bracket faces backwards, the jack will not lie flush against the wall unless it is in an outlet box.

(2) Turn the contact rails so that they do not accumulate dust and debris. (Sometimes the directions show them mounted like Fig. 8-21. This is incorrect. Mount them like Fig. 8-22, instead.)

Mounting

Outlet

(1) Attach the bracket to the outlet with screws in the prepared holes.

(2) If there is no bracket, terminate first. Then attach the jack to the outlet box with screws in the prepared holes.

No Outlet

(1) Use a folding ruler to place the jack level with other outlet boxes (electric).

Fig. 8-21. Flush jack upside down—how *not* to do it.

(2) Cut a hole in the drywall with a 2-1/4 inch hole saw.
(3) Pull wire through the hole.
(4) Use a plumb bob to align properly.
(5) Mark where you will insert the wall anchors.

Fig. 8-22. Flush jack rightside up—how to do it.

(6) Insert wall anchors.

(7) If there is a bracket, attach it. If there is no bracket, terminate, then mount block.

Termination

If the jack has screw terminals:

(1) Remove 4 inches of sheath from the station wire.

(2) Strip 3/4 inch of insulation from the conductors.

(3) Make a hair pin turn with the stripped wires around the corresponding color terminals. Some will be between washers (one wire per space between). Do not pull up too tight. Snug the nut down gently.

(4) If the wire has been run continuously (see Chapter 7) you can terminate without cutting. If the wire is not run continuously or even if a cut occurs accidentally to a continuous wire, terminate both ends.

Cover

(1) Screw the connecting block into the bracket.

(2) Screw the cover onto the block.

If the jack has split beam terminals (Fig. 8-23):

(1) Remove 4 inches of sheath from the station wire.

(2) Do not remove insulation. Note: 19-gauge wire is very stiff. If your wire is 19 gauge, you will have to strip the insulation to allow the split beam to work.

Fig. 8-23. Closeup view of split beam terminations of flush jack.

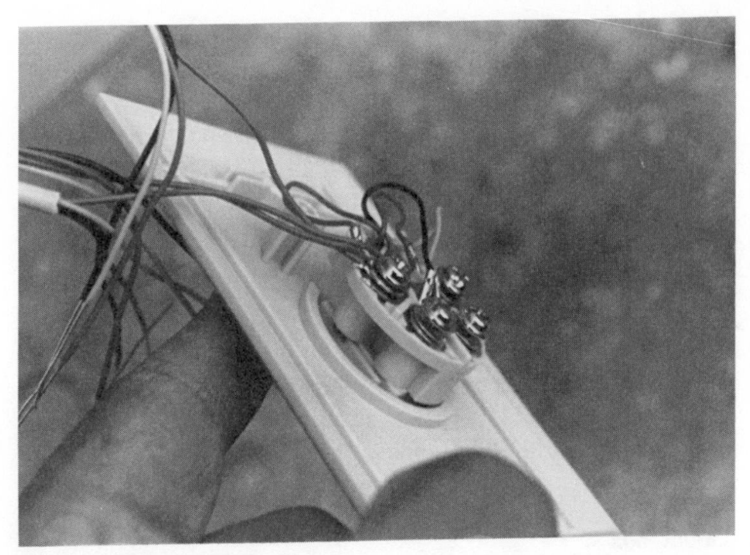

Fig. 8-24. Splice of flush wall jack.

(3) If the wire is run continuously (see Chapter 7) you must still cut it in order to terminate, but only cut the four wires that you will be using. Leave the other wires untouched.

(4) Unscrew the terminals two and one-half times and insert the wire into the ports. Tighten the screws.

Installing Cover:

Screw the connecting block into the bracket. A flush wall jack can be used as a splice. See Fig. 8-24.

WALL PHONE (FIG. 8-25)

Positioning

(1) Allow room for the actual phone beyond the dimensions of the jack.

(2) Turn the contact rails so that they do not collect dust and debris.

Mounting

(1) If the surface to which you plan to mount the jack is flimsy, locate a stud.

(2) Use a plumb bob to assure correct alignment.
(3) To mount block

□ If the wood is not likely to crack you can:

(1) Attach a hex head screw 1 inch with a 1/4-inch nut driver
(2) Finish with an electrician's screwdriver.

□ If the wood is brittle you must:

(1) Drill the holes with a 3/32-drill bit.

(2) Attach with 1-inch pan head screws. It is possible to mount a wall phone one of three ways: in an outlet box, on a stud, or in dry wall with wall Molly® bolts.

Outlet Box

(1) Pull wire through outlet box.

(2) If it has screw terminals, remove four inches of sheath (six inches if the wire is run continuously).

(3) Pull the wire through the corresponding color openings in the block.

(4) Screw the block into the outlet box.

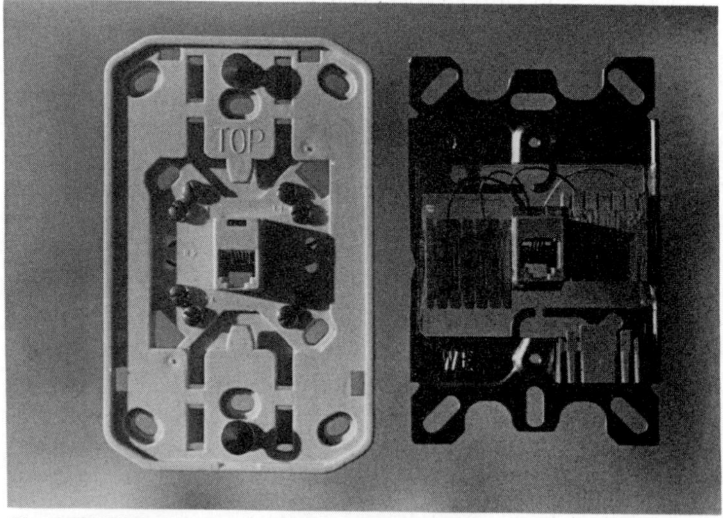

Fig. 8-25. Wall phone jacks without covers.

On Stud

(1) Sound the wall to find a stud (hollow is not stud, dead sound is stud).

(2) Measure with a folding ruler to place at a height that is like other outlets (electrical or phone). Fifty inches to 52 inches from the floor to the bottom of the jack is usual.

(3) Align with a plumb bob.

(4) Screw in the block.

(5) Staple wire next to the edge of the jack. Wire is usually brought in from underneath.

Dry Wall With Wall Anchors

Wire can be run from inside (like an outlet), or from outside (like a stud).

(1) Measure with a folding ruler to place at the height of other outlets (electric or phone). Fifty to 52 inches is the usual.

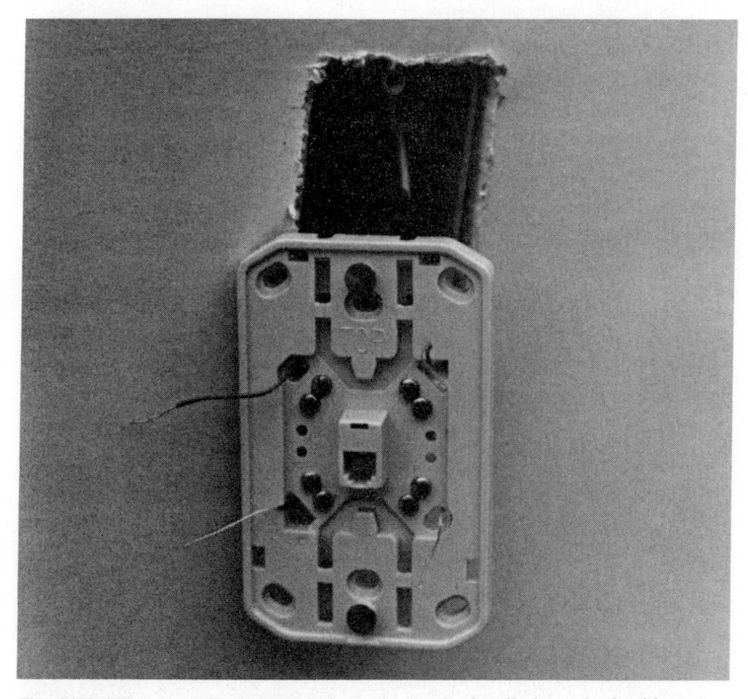

Fig. 8-26. Wires entering properly from the back of wall phone jack with screw terminations.

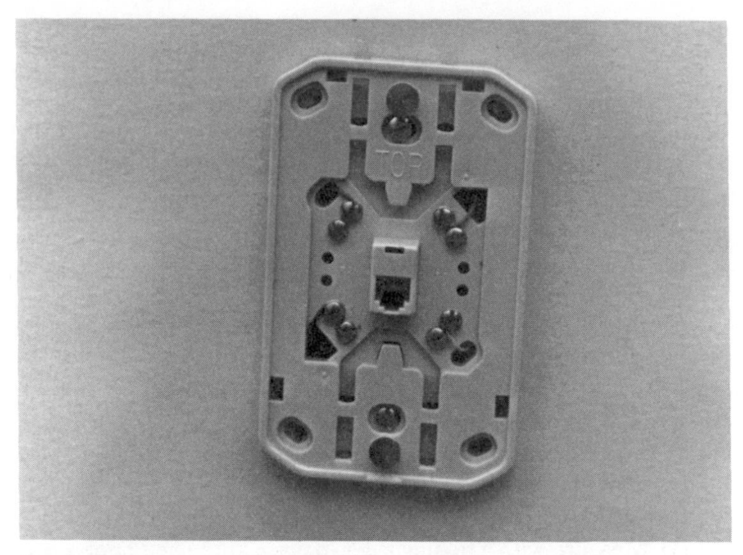

Fig. 8-27. Wall phone jack mounted and terminated in outlet.

(2) Align with a plumb bob.
(3) Mark the location for the Molly® bolts.
(4) Insert the wall anchors.
(5) Terminate if you are running the wire from the inside.
(6) Attach the jack.

Termination

(1) SCREW TERMINALS

□ Remove 4 inches of the sheath. Six inches if the wire is run continuously—see Chapter 7.

□ Strip 3/4 inch insulation (Fig. 8-26).

□ If the wire has been run continuously, you can terminate without cutting. If the wire is not run continuously or a continuous wire is cut accidentally, terminate both ends on the same color.

□ Make a hair pin turn with the stripped wires around the corresponding color terminals. Some will be between washers (one wire per space between). Do not pull up too tight. Snug the screws down gently (Fig. 8-27).

(2) SPLIT BEAM STRIP

□ Remove four inches sheath.

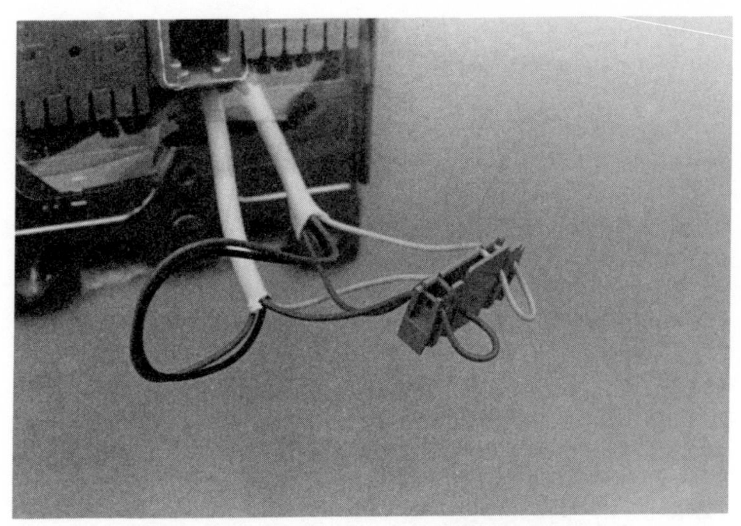

Fig. 8-28. Looping wires into wall phone jack with connector cap.

☐ Do not remove insulation unless it is stiff #19 gauge wire.
☐ If the jack has:

Connector Caps (Fig. 8-28)

(1) Loop wires through cap.

Fig. 8-29. Pushing connector cap up with screwdriver.

(2) Force the cap up into the slots of the terminal strip with a screwdriver (flat blade type) until seated (Fig. 8-29).

No Connector Caps

(1) Loop wires, if continuous (see Chapter 7 for instructions on stripping the sheath), if not, bring both ends inside the corresponding colors at the bottom of the terminal strip.

(2) Force the wire into the terminations until it is seated. The tool for this is the lead insertion tool or you can use a small (1/8-inch) bladed screwdriver (Fig. 8-30).

☐ Fold the wires flat. Be sure that one does not cross over another. Fold down the plastic flaps (Fig. 8-31).

Cover

Screw Termination

The cover just snaps on (Fig. 8-32.).

Split Beam

Screw the cover onto the bracket. On some of these types you also need to screw on phone mounting lugs (furnished with jack).

If you use this jack to splice:

☐ Split beam—This is just like installation except you will

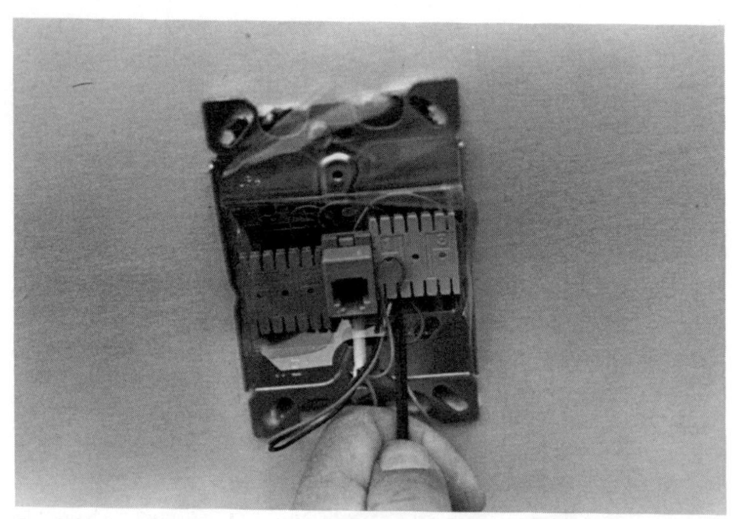

Fig. 8-30. Terminating wires with small bladed screwdriver into split beam strip.

Fig. 8-31. Plastic flaps folded over properly terminated looped wire.

need the unused breakout tabs and the unused ports.

☐ Screw—Loosen screws. Move old wire to bottom of screw (below washer next to connecting block), connect new wire where old was.

Outside Jacks

Do not span the wire in the air. Run the wire in conduit when possible. Mount the jack in the driest area possible (Fig. 8-33).

Wire Junction

This is only for splicing (Fig. 8-34). The wires are made down and can be plugged and unplugged like a modular jack. Install it just like the spring type jack.

Round Flush Jack

This is installed just like other flush jacks except (1) you cut a hole in the drywall with a 1-3/4 inch hole saw (2) place the bracket

Fig. 8-32. Wall phone jack with cover.

Fig. 8-33. Outdoor modular jacks with protective housings.

on the hole. Use a plumb bob and line the screw holes up perpendicular to the floor (3) mount the bracket with wall anchors.

DO'S AND DON'TS OF JACK INSTALLATION

These simple tips may help prevent problems with your telephone jacks.

1. Mount jacks so that debris cannot collect on the contact rails in the modular connection (where the phone plugs in).
2. Always attach a jack to a firm surface where there is no play. If the wall or baseboard is not sturdy, mount the jack on a stud.

Fig. 8-34. Wire junctions (to with modular cord, bottom without).

3. Never mount a jack in a damp or humid area. If you must, enclose it in a weatherproof outlet box.

4. When connecting conductors to the jack terminals, do not allow excessive copper (shiners) to be exposed. This will cause trouble on the phone line if the copper touches other parts of the jack. (No more than 1/8-inch exposed copper).

5. The split beam termination system is designed so that only one wire is to be connected at a time. A second termination is provided to allow a second wire.

6. In a screw termination, insert a wire between each washer.

7. Do not twist wires together before terminating on a screw terminal.

8. Be careful not to pinch or cut the conductor wires when you screw the cover on.

9. When you need to bridge wires, use an acceptable connector such as a jack or wire junction.

10. Do not try to terminate stranded wire or solder wire into telephone jacks.

11. Connect wires on screw terminations clockwise and with a hairpin turn to keep the conductors from getting caught between the threads and washer.

12. Mount jacks away from areas where they are susceptible to damaging insect sprays and rug fibers. If this is unavoidable, use the humidity and dust proof covered jack mentioned earlier in this chapter.

13. Leave slack wire so that you can manipulate equipment easily and, if you make a mistake, reterminate. Wire does not grow!

Chapter 9

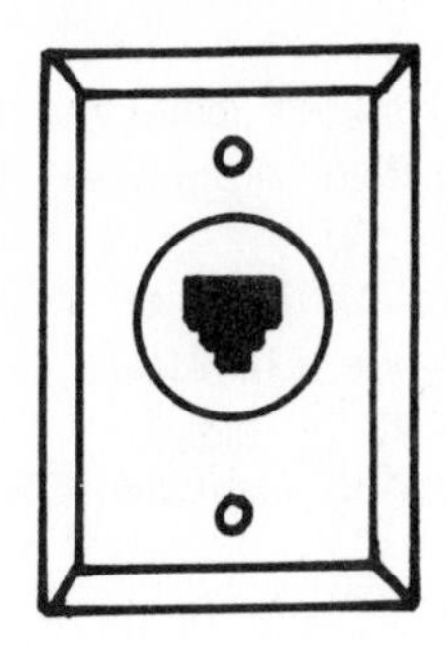

Tools

THE TOOLS REFERRED TO THROUGHOUT THIS BOOK ARE listed below and their use is explained.

- ☐ Screwdriver (electrician's 4-inch to 6-inch with a 1/4-inch blade with an insulated handle)
- ☐ Screwdriver (4-inch length with a 1/8-inch blade)
- ☐ Diagonal Cutting Pliers (5-1/2 inch to 6-inch)
- ☐ Longnose Pliers (5-1/2 inch to 6 inch)
- ☐ Push Drill (manual)
- ☐ Nut Driver (1/4-inch (hex-head))
- ☐ Wire Stripper
- ☐ Staple Gun (Arrow) (T-18 or T-25 or T-50)
- ☐ Fishtape
- ☐ Power Drill (3/8-inch heavy duty)
- ☐ Bell Hanger Drill (1/4-inch in diameter length 12- or 18-inches)
- ☐ Hole Saw (2-1/4 inch or 1-3/4 inch) for power drill
- ☐ Key Hole Saw
- ☐ Folding Ruler or Yardstick
- ☐ Tape Measure
- ☐ Scotchlok® Tool
- ☐ Homemade Plumb Bob
- ☐ Claw Hammer
- ☐ Tack Hammer

☐ Spade or Auger Bit (1/2-inch or 5/8-inch for power drill)
☐ Brace and Bit
☐ Tone Generator for tracing wire (gives a tone to be heard on wires)
☐ Testset made from telephone and adapter
☐ Sectioned Cane Pole or Zip Pole
☐ Coat Hanger
☐ Socket (3/8-inch) and Handle
☐ Utility Light, Lantern or Flashlight

What to Look for and How to Use

☐ Screwdriver—The blade should be squared, not sharp like a knife, so that it will fit the slot of the screw properly. The handle needs to be large enough to give good torque and looks like it is not going to come loose. Having it insulated is not necessary but does give more torque because of the larger grip. The plastic handle of the screwdriver in Fig. 9-1 is not insulated but the rubber covered handle of the screwdriver (speedriver) is insulated. The small (1/8-inch blade) screwdriver can be substituted for the tool made to terminate certain split-beam jacks.

☐ Diagonal Cutting Pliers—they should have plastic handles. The cutting edges should meet each other so that when closed no light shows in between the cutting edges. They are used to cut and strip the sheath from the station wire. With practice, they can also be used to strip insulation from the conductors.

☐ Longnose Pliers—the handles should be covered with a plastic whose grasping surfaces are rough enough to file a finger nail. They are used to handle the bare conductor wires around screw

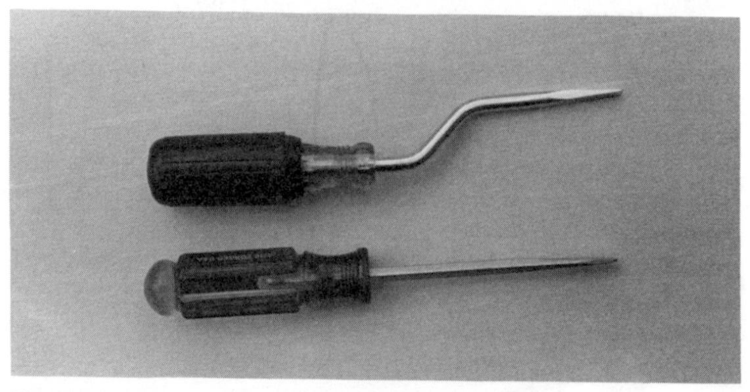

Fig. 9-1. Insulated screwdriver with standard screwdriver.

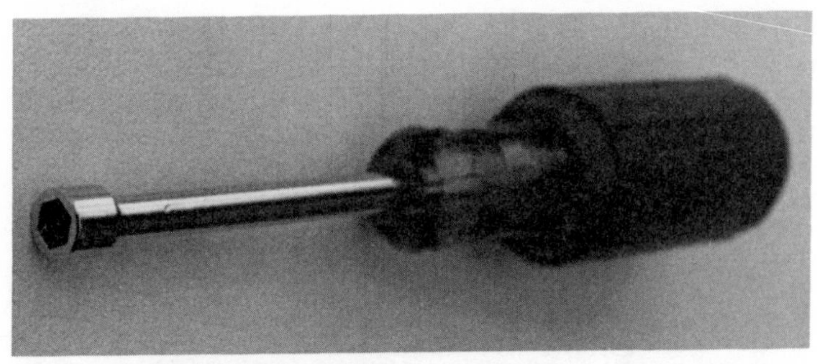

Fig. 9-2. Insulated 1/4-inch nut driver.

terminations and to pull the cord that peels the sheath from the station wire.

☐ Push Drill (optional)—Follow the directions for proper care that come with it. The bits do not last long, so if you are going to use it extensively, get several bits in the needed size. Use this drill to make starter holes (for pan head screws) or in wood that is likely to split.

☐ Nut Driver (1/4-inch)—This driver works with the 1/4-inch hex-head screws. When installing some jacks it may be necessary to tighten the screw with a screwdriver to make it flush (Fig. 9-2).

☐ Wirestripper—This tool comes in many styles and price ranges but the simple ones work the best for installing. Practice with scrap wire.

☐ Staple Gun—Figure 9-3 shows the brand I use, Arrow (T-18 or T-25 types). These are not easy to find at the moment, but people doing their own installation will increase the market. The T-18 is the better of the two for stapling four conductor wires, which are never larger than 3/16-inch in diameter. The T-25 works well with larger wire (six, eight, or twelve conductor). I prefer the 3/8-inch staples. They drive in more easily and hold more tightly on most wood surfaces.

☐ Fishtape (optional)—It is used to pull wire through conduit, to sneak wire around walls or ceilings, and in some instances to fish for wire.

☐ Power Drill (3/8 inch)—The drill will need to be heavy duty or commercial so that it can take the strain of drilling through a wall or other solid structure. It should either have *Double Insulated* printed on the side or have three prongs. See Chapter 1 for more details.

☐ Bell Hanger Bit (1/4 inch in diameter and 12 or 18 inches

Fig. 9-3. T-18® staple gun.

in length)—These are available for wood or masonry. The masonry bit has a carbide tip. *Do not* use the wood bit on masonry or the masonry bit on wood. If you are drilling through a wall that has wood, then masonry, use the wood bit, then the masonry bit. The bell hanger bit allows you to pull back the wire through the hole you have drilled. Whenever you drill through an outside wall, drill from the inside out. Drill slightly downward to prevent moisture leaking in. Run the drill bit back and forth several times to smooth and enlarge the hole so that the bit and wire will come back through smoothly. Use this tool as shown in Figs. 9-4 through 9-10.

At the top of Fig. 9-4 is a wood bit, on the bottom is a masonry bit. Figure 9-5 shows the hole in the bit through which wire will go. Unsheath about 6 inches of wire as shown in Fig. 9-6. Figure

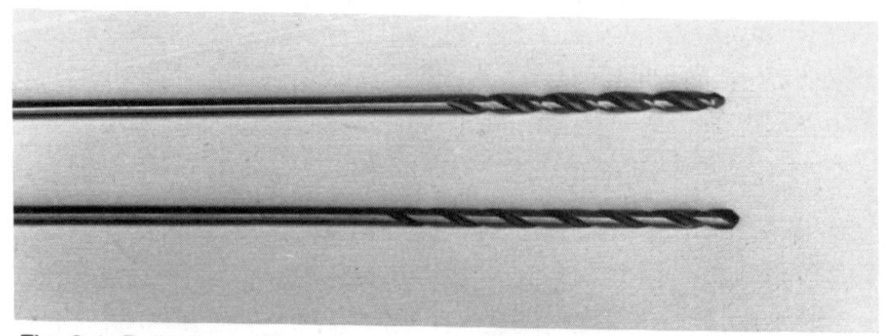

Fig. 9-4. Bell hanger bit, top wooden, bottom masonry.

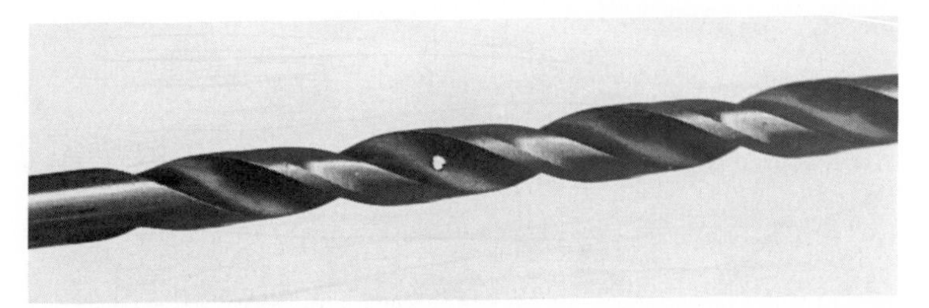

Fig. 9-5. Hole through shaft of bell hanger bit.

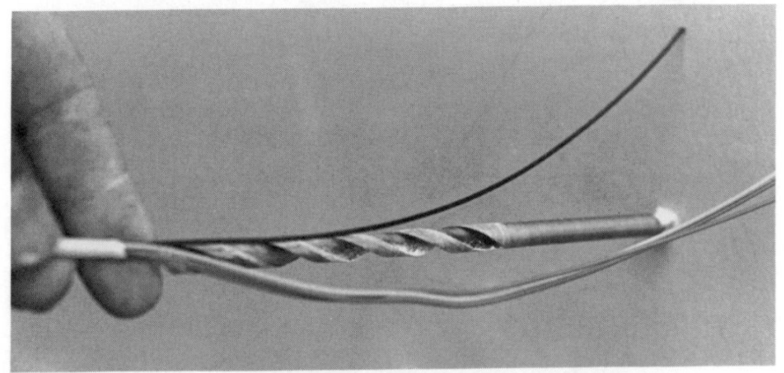

Fig. 9-6. Station wire stripped of sheath, showing conductors.

Fig. 9-7. Single wire without insulation.

9-7 shows where the sheath ends, cut off three of the wires (leaving one). Strip that one of insulation. Insert the stripped wire in the hole (Fig. 9-8). Match the sheath to the tip of the bit. Wrap the wire in one of the twisted channels, back toward the sheath.

Fig. 9-8. Wire inserted in hole and wrapped in channel.

Fig. 9-9. Wire wrapped to other side of channel and wrapped.

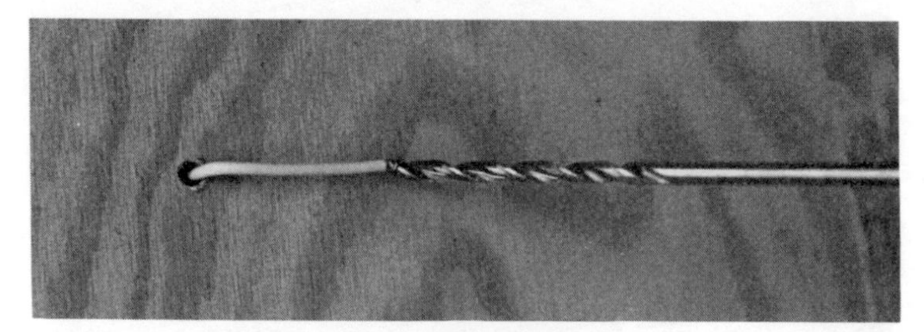
Fig. 9-10. Wire pulled through to other side of wall.

Wrap the rest of the stripped wire into the opposite channel that also runs back toward the sheath (Fig. 9-9). Wrap the wire around the tip of the bit about three times. Trim the excess wire. Pull the wire through the wall as shown in Fig. 9-10.

When drilling be sure the bit slides in and out smoothly or else the bit will be hard to pull with the wire attached. When drilling out an outside wall, drill slightly downward; this is to help prevent moisture from coming in.

☐ Spade or Auger Bit (1/2- or 5/8-inch)—If you are prewiring a new home, the wire has to be run between top plates or studs. An electric drill speeds up the job.

☐ Hole Saw (2-1/4 inch or 1-3/4 inch)—This saw attaches to a power drill to cut a hole in drywall (or other wall surfaces) easily. The hole that it makes allows the use of certain flush jacks that do not need outlet boxes.

☐ Keyhole Saw—This is a hand saw that works very well to cut holes that must be large enough to mount an outlet box (or any hole size for termination of a flush jack).

☐ Folding Ruler or Yardstick—Use this to measure a short distance, such as the height of outlet boxes or the height of a wall phone jack.

☐ Tape Measure—This is used to measure long distances to determine the amount of wire needed to prewire or rough-in a home.

☐ Brace and Bit—If there are places where you cannot use a power drill, use a brace and bit. It will work in most situations but is often slow. Be sure the bit is sharp before you start. Bell hanger bits are available for braces.

☐ Scotchlok® Tool—This tool is made especially for the use of Scotchlok® connectors. It can also be used to cut the wires that you plan to splice. A pair of 6-inch slip joint pliers can be substituted.

☐ Homemade Plumb Bob—This tool is easy to make. All that is needed is a length of string and a weight (such as a key). It will help you to line up brackets for flush jacks or wall outlet boxes. It may even be used to determine where wire is in a wall.

☐ Tone Generator—This electronic device is used to trace wires in a home to determine where there is lack of continuity (an open) or a short. It is used in conjunction with the homemade test set which allows a fluctuating tone to be heard over the wires. Currently, it is not readily available to the public, but with the changes in the telecommunications industry it may soon be.

☐ Claw Hammer—All that is needed to install the simplest (two nail) outlet box is a hammer. Just measure the height and drive the nails into the stud.

☐ Tack Hammer—If you decide not to use a staple gun, insulated staples are attached with a tack hammer.

☐ Testset—The most important tool to any person doing telephone work, the testset, makes it easy to check for dial tone

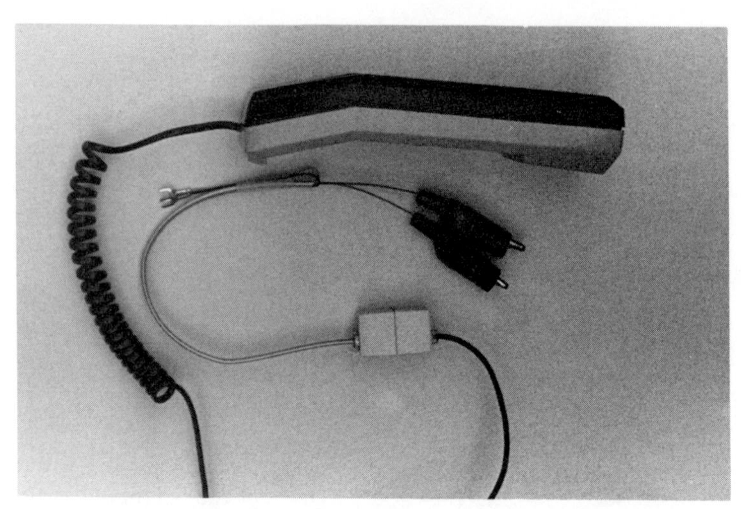

Fig. 9-11. Test set made with phone in-line coupler and alligator clips.

or other problems on the line (Fig. 9-11). There are many ways to make a test set. The least expensive way is to: (1) purchase a disposable phone or one that is just a handset with no base (2) an in-line coupler (3) a base cord, which is a mounting cord with spade tips on one end and a modular plug on the other, and (4) two alligator clips. (5) Attach the alligator clips to the green and red tips of the base cord. (6) Tape or cut off the yellow and black leads (they will not be used in testing). (6) Plug the cords together and tape the cords at whatever length you want them. Voila! You now have a testset!

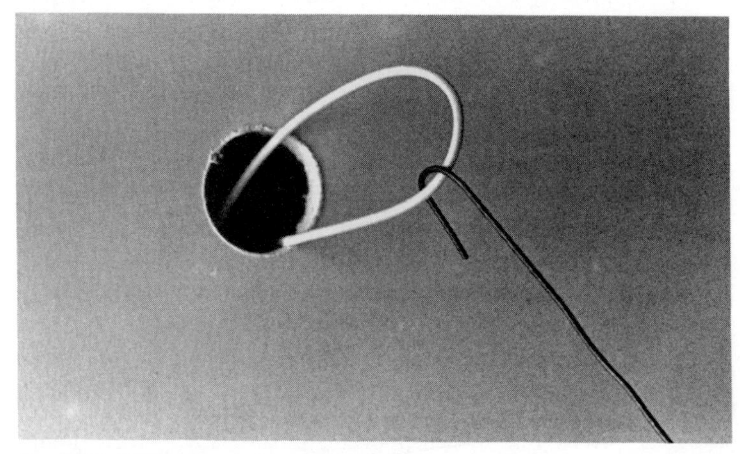

Fig. 9-12. Coathanger being used to fish wire out of wall.

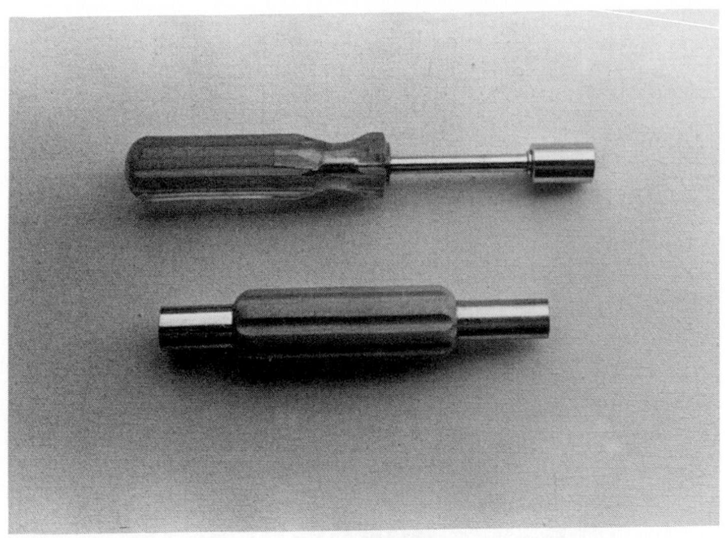

Fig. 9-13. 3/8-inch socket and handle with can wrench below.

☐ Sectioned Cane Pole or Zip Pole—There may be occasions when you are not able to reach a wire in the attic or crawl space under a home. Tape a hook to the end of a cane pole or zip pole. Push plenty of wire into the area you cannot reach. Hook it with this tool.

☐ Coat Hanger—A coat hanger can be used to fish a wire from in the wall as shown in Fig. 9-12. Cut the hanger with the diagonal cutting pliers to the length desired and hook the end.

☐ 3/8-Inch Socket and Handle—There may be times when it is essential to remove wires from the outside protector (see Chapter 11). This tool will work on all protector nuts that terminate the inside wiring (Fig. 9-13). Remember, if you have an interface you should not be touching this protector.

☐ Utility Light, Lantern, or Flashlight—There is usually little or no light in the attic or crawl space under the home. It is important to have a good light source. Be safe. Use a dependable light.

Chapter 10

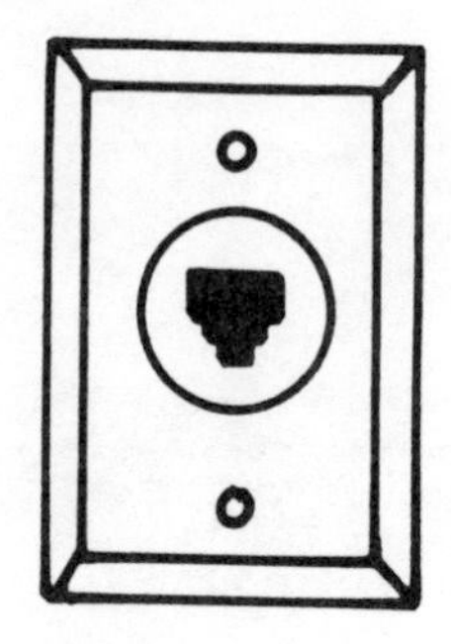

Equipment

BELOW IS A LIST OF THE EQUIPMENT REFERRED TO throughout this book. The specific job will state which equipment is needed. Once you know what equipment you need for your task, you can refer to this chapter for a description and information on the use of that equipment.

☐ Screws—Pan head or Slotted Hex head (8 × 3/4-inch or 1-inch)
☐ B Wall Anchors size 4. Fig. 10-1 shows pan head, hex head, and wall anchors.
☐ Silicone (clear) Sealant
☐ Electrical Vinyl Tape (black)
☐ Scotchlok Connectors (UR or UY type)
☐ Cable Ties or Tie Wraps
☐ Staples (Arrow) T-18 or T-25 (3/8-inch)
☐ Insulated Staples
☐ Plastic Anchors
☐ Outlet Boxes
☐ Romex Staples

Description and Use

☐ Screws—Pan head or hex head screws need to be attached to surfaces that do not give. This means you should not use them

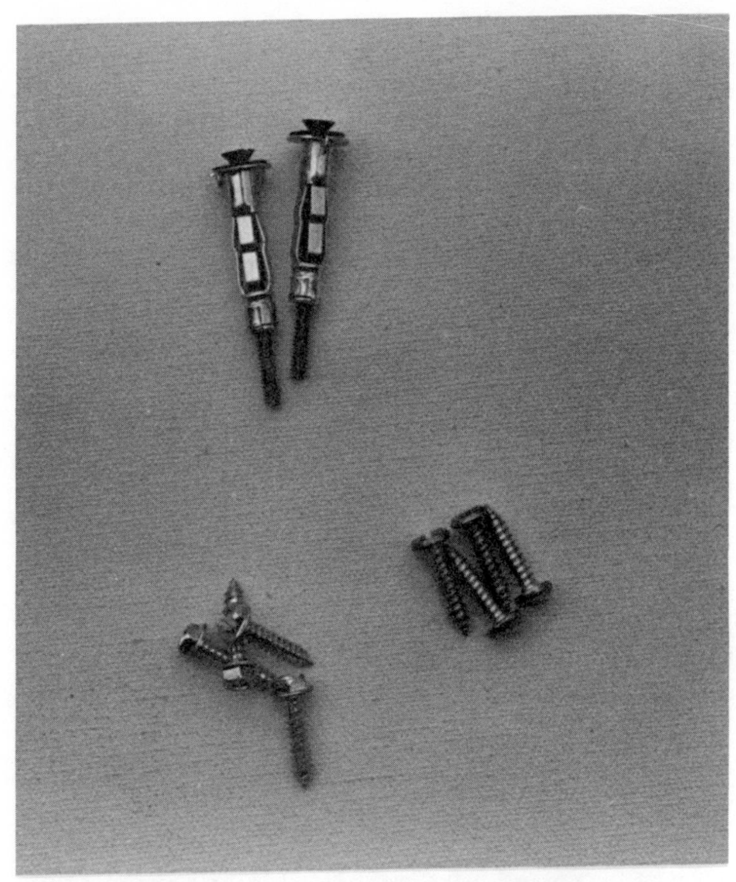

Fig. 10-1. Pan-head and hex-head screws and Molly® bolts.

directly on drywall. Drill holes with a 3/32-inch bit. To hold a jack, if you screw directly into wood, use 3/4-inch screws, but if the screw has to go through the wall cover to get to the wood or other material, use 1-inch or longer screws. If the material is not hard and brittle so that it will not crack, you will not have to drill a hole. You can use slotted hex-head screws, installed with a 1/4-inch nut driver.

☐ B Wall Anchors (Molly® bolts) (#4 or size 1/2-inch)—Use these to mount anything flush, especially on drywall. The sizes given work for 1/2-inch (or less) thick drywall. If the wall is thicker, use a larger anchor.

☐ Silicone Sealant (clear)—Sealant is used to plug up holes that are bored through the outside wall of a home. Apply last, after the wire is pulled and fastened.

☐ Electrical Vinyl Tape (black)—This is used to help pull wire or tape certain splices.

☐ Scotchlok® (3M) Connectors (UR or UY type)—Fig. 10-2. If you are going to splice without the use of a connecting block or a wire junction, do it with a product made for the job. UR and UY are just two of several types of connectors made for this. The UR type allows three wires to be joined at one time. The UY type only allows two.

(1) Remove 3/4 inch of the sheath. The wires do not need to be stripped of insulation for proper contact.

(2) Give the wires one full twist.

(3) Cut them even.

(4) Push with the button in the connector down, so that you can see that the wires are all the way into the conductor ports.

(5) Crimp the connector. There is a tool made to crimp the connectors to the wires but you can use a pair of slip joint pliers. These connectors contain a sealant that prevents the wires from corrosion where the splice is made. (Caution: The sealant is irritating to the eyes. In case of contact, flush with water and contact a physician.)

☐ Cable Ties or Tie Wraps—These are used when you need

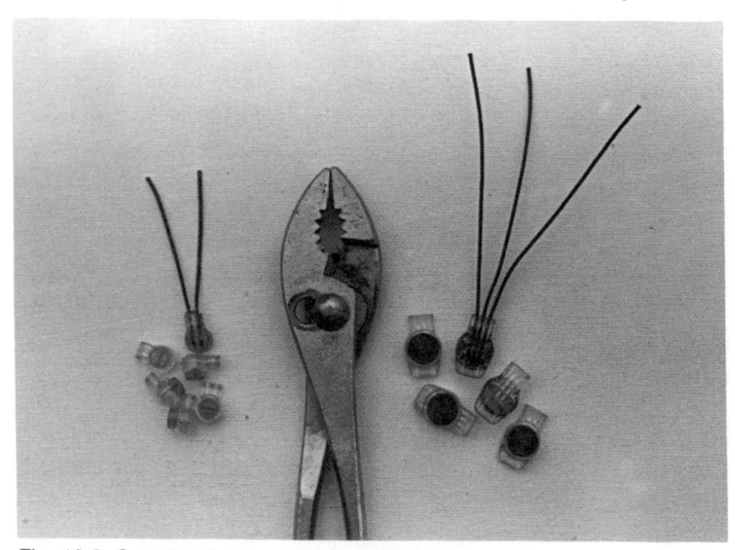

Fig. 10-2. Scotchlok® UR and UY connectors.

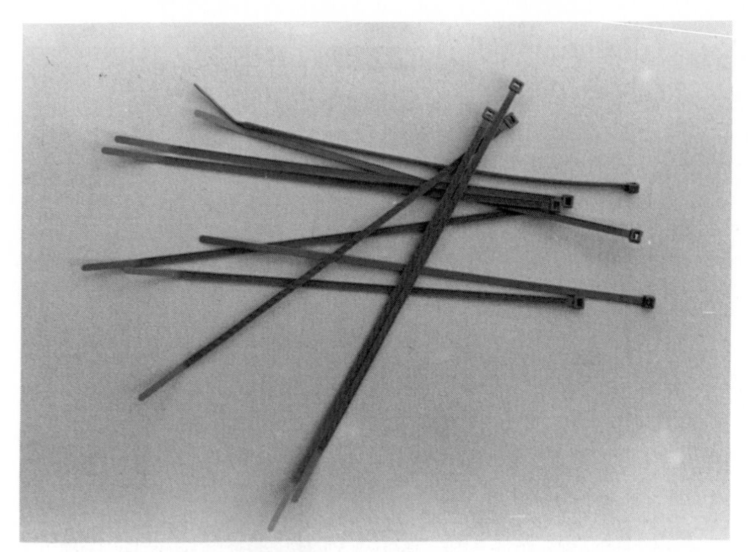

Fig. 10-3. Cable ties or tie wraps.

to run wire in places where you cannot staple (Fig. 10-3). Tie them to conduit or other permanent fixtures. Cut off the excess with diagonal cutting pliers. There are also some cable ties that have screw holes to help mount on flat surfaces. Gray or white should only be used indoors because they will disintegrate outside in ultraviolet light. The best type to use outside are carbon black cable ties. When possible, tie the wires to the back of conduit or other attachments. The wire will be less likely to be cut and will be less obtrusive.

☐ Staples (T-18 or T-25) 3/8-inch ARROW—T-18 is most commonly used for four conductor wire or quad. T-25 is for larger wire (six conductor to twelve conductor). Staples that are 9/16-inch work fine if the surface is not very hard. Do not staple directly to plaster (this includes drywall).

☐ Insulated Staples—These staples come in a variety of sizes to suit different wire. You do not need a staple gun, as these are fastened with a hammer. Run the wire the same way you would if you were using a staple gun.

☐ Plastic Anchors—These are used only on surfaces that are solid and hard. They should not be used on drywall. My experience is that they work well on masonry. Make the hole slightly smaller than needed, then gradually enlarge by reaming with the bit. The plastic anchor needs to be forced in some so that it can grip as it is being expanded by the screw.

Fig. 10-4. Outlet box (side clamp).

☐ Outlet Boxes—There are a variety of outlet boxes from which to choose. If you are prewiring, use the type with two nails that drive into the stud with a hammer. For flush jacks with an outlet box, installed after the walls are up, use a side-clamp box (Fig. 10-4) or those with spring clips unless the walls are flimsy and you have to mount on a stud.

☐ Romex Staples—These usually hold electrical sheathed cable or flexible metallic conduit. Our application is to hold telephone wire out of the way. They are hammered in before the wire is pulled through.

Chapter 11

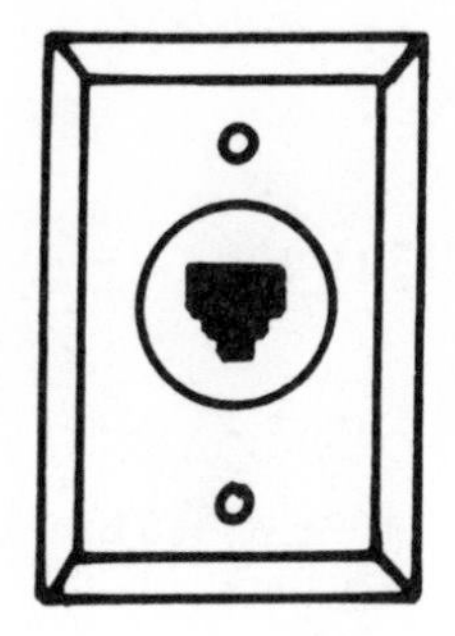

Troubleshooting

NO MATTER WHO INSTALLS YOUR TELEPHONE SYSTEM, NO matter how well it is installed, sometime, somewhere, something is going to go wrong! You will have "trouble" on your line. This chapter will help you to locate the trouble and determine how to correct it. Sometimes you will need to contact the telephone company, sometimes you will replace a part, and sometimes you will repair something. If you can correct the problem yourself, that is to your advantage. Even if you cannot, you can save money by isolating the specific problem before you get help.

This chapter is organized so that you first decide what trouble you are having. You read beneath that heading to determine whether the trouble is located in the telephone line, inside wiring, a jack, a cord, or the phone itself. Once you have located the difficulty, you determine the physical cause and its solution.

These four terms will be used in the following directions to identify types of trouble you may have: open, short, ground, and cross. An open is a break in the conductor wire and looks like Fig. 11-1. A short occurs when two conductors on the same circuit are touching one another. It looks like Fig. 11-2. A ground occurs when a conductor is making contact with a ground source, such as a pipe or water. It may look like Fig. 11-3. A cross occurs when a conductor wire is touching a wire in another circuit. It will look like Fig. 11-4.

Many of the problems encountered will be in jacks, on wire,

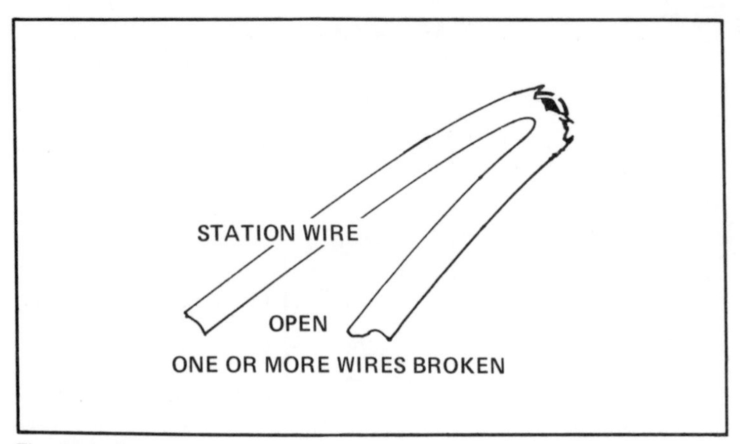

Fig. 11-1. An open.

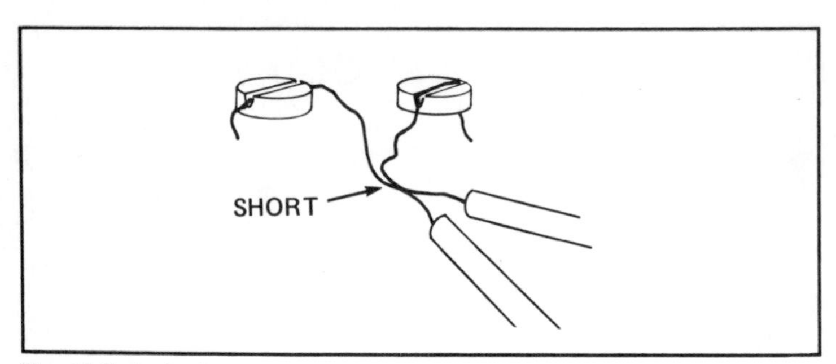

Fig. 11-2. A short.

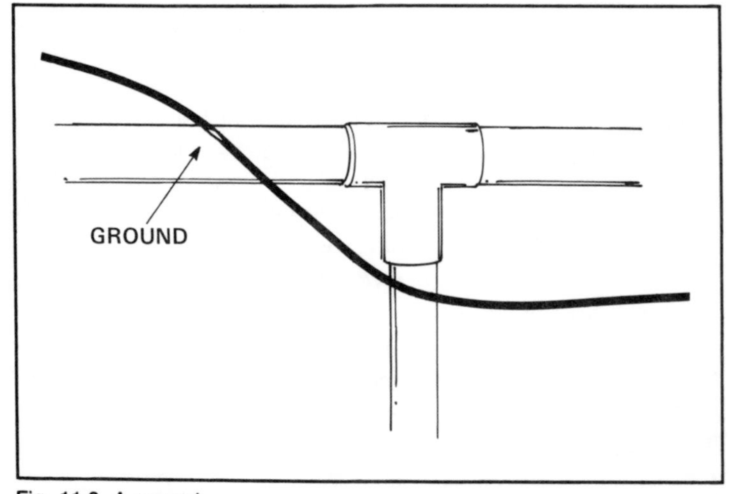

Fig. 11-3. A ground.

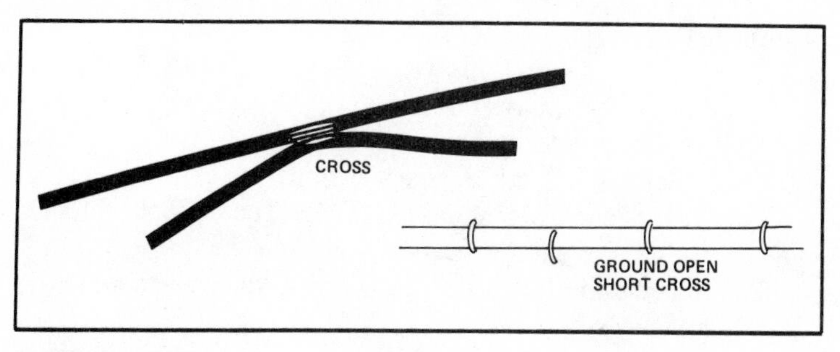

Fig. 11-4. A cross.

etc., but occasionally the trouble will be in the telephone. Refer to the chapter on phones for the legalities of working on your own phone. The information in this chapter shows you what problems you may encounter in standard and electronic telephones. If you tamper with a telephone, the warranty may become void. If you have difficulty finding a repair person for a broken telephone, the FCC has a list of refurbishers who will repair your telephone or provide reconditioned telephones. Send a written request to:

Federal Communications Commission
Consumer Assistance Office
Room 258
Washington, D.C. 20554
(202) 632-7000

IDENTIFY THE PROBLEM

(1) No Dial Tone
(2) "Goes dead" or disconnects
(3) Noises on Line

- (a) Humming
- (b) Other Voices
- (c) Static
- (d) Radio Signals

(4) Bell

- (a) Does not ring
- (b) One short ring
- (c) Soft ring

(5) Dialing

(a) Dial Tone stays on
(b) Dials wrong numbers
(c) Bell taps or clicks when Dialing
(d) Does not Dial Out or No DTMF (Touch-Tone™) tones

(6) Transmission (Receiving or sending voice signals)

(a) Can hear caller but caller cannot hear you
(b) Cannot hear caller but caller can hear you
(c) Other party hears your voice distorted
(d) voice of either party decreases in strength

NO DIAL TONE

I Test the Phone

(A) Look to see that the telephone is completely plugged in at all connections.

(B) If you have an extra working phone, plug it into the jack.

☐ If that phone works then your trouble is in the first phone. You should:

(a) replace phone or swap out damaged cord
(b) swap parts
(c) repair (Send to manufacturer or refurbisher for repair)

☐ If the second phone does not work skip to II

(C) If you do not have an extra phone, plug the nonworking phone into a working jack. If none of yours are working, use a neighbor's.

☐ If it still does not work, the trouble is the phone.

(a) replace phone or swap out damaged cord
(b) swap parts
(c) repair (Send to manufacturer or refurbisher for repair)

☐ If it does work, see II

II Test the Line Coming to the House

Note: If there is only one bad jack, you might physically inspect it (see IV. "If Your Trouble is an Open"). If that does not work, return to this point. Determine whether your home has an interface or a protector. Check Chapter 2 for more information.

Interface

(A) Locate Network Interface

(B) Unplug the modular connection to the interface.

(C) Wait five minutes for the line to clear, then plug the telephone into the interface. Do you get dial tone?

☐ If you do not get dial tone, the problem is with the telephone company. Unplug your phone. Plug the modular connection back in. Call the phone company for repair.

☐ If you do get dial tone, the problem is in the inside wiring or jacks. Go to III.

Protector

(A) Locate the protector.

(B) Remove the cover.

(C) Inspect the wires on the terminals of the protector.

☐ If corroded—spray with WD40 and work nuts in and out.

☐ If loose—tighten nuts

☐ If breaks or open—reterminate

(D) Connect a homemade test set (see Chapter 9) to the protector terminals as shown in Fig. 11-5.

☐ If you get dial tone, the problem is an open (break or cut). Go to IV.

☐ If you do not get dial tone, disconnect the testset. Remove all inside wiring from the protector terminals as shown in Fig. 11-6 (use a 3/8-inch socket on handle). Keep track of which wire comes from which terminal. Reconnect the testset.

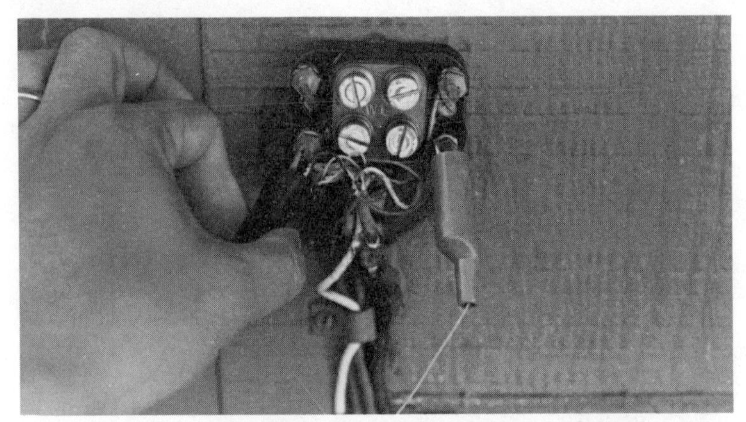

Fig. 11-5. How to connect the alligator clips of a testset to protector terminals.

(a) If you still do not get dial tone, call the phone company for repair. Disconnect the testset and reconnect the wiring.

(b) If you do get dial tone, the problem is a short. Go to III—Protector.

III Test the Inside Wiring

Interface

(A) Remove the telephone and replug the modular cord into the interface.

Fig. 11-6. Removing wires from protector terminals to check each matching pair.

(B) Remove the cover from the wire junction or from the jack (to which all of the inside wiring is terminated).

(C) Clip the testset (see Chapter 9) onto the red and green terminals in the wire junction (or jack).

- ☐ If you still get dial tone, there is an open (break or cut) in the wire in the wire junction, in a jack, or in the wire between them. Go to D.
- ☐ If you do not get dial tone, the wire is making contact with itself, causing a short. Go to D.

(D) Physically examine the wires in the junction to see if any are bare (shiner) or broken.

- ☐ If the wires are damaged, reterminate them.
- ☐ If the wires are not damaged, go to IV.

Protector

The inside wiring should still be disconnected from the terminals, and the testset still connected to the terminals of the protector.

(A) If you have a short, place one pair of matching wires at a time on the terminals as shown in Fig. 11-7.

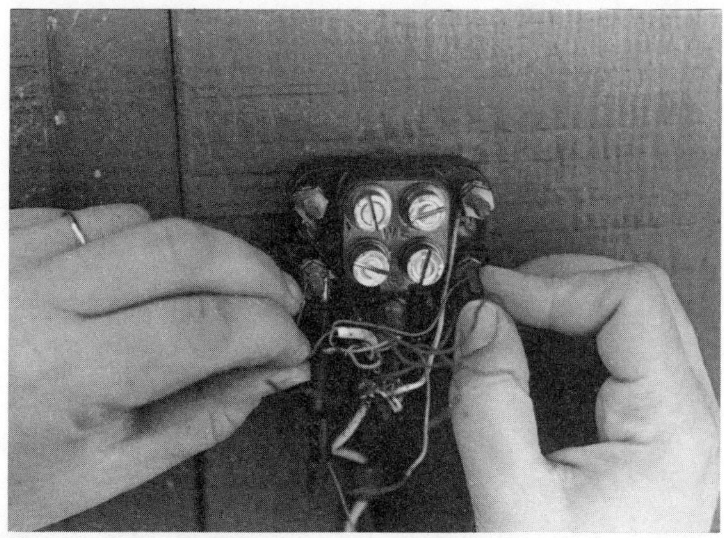

Fig. 11-7. Place matching wires on the terminals as test set is connected.

☐ If the pair you test allows dial tone, go on to the next pair.

☐ If the pair you test does not allow dial tone, the wire is making contact with itself and creating a short.

(a) Inspect what is visible of the wire for shiners (bare wire).

(1) If you find the damage, repair and reterminate the wire to the protector, clearing the problem.

(2) If you cannot see the damage, only reterminate the good wires. Go to IV.

(B) Reterminate the good wires to the protector. Go to IV—SHORT.

IV Inspect Inside Jacks

If Your Trouble Is an Open:

(A) Select which jack to start with.

☐ If one jack is not working, inspect it.

☐ If more than one jack is not working, begin with the one that is closest to the protector or wire junction.

(B) Remove the jack.

(C) Clip the homemade testset to the tip and ring (red and green) terminals.

☐ If there is dial tone, the problem is in the jack. Visually inspect the jack for corrosion, breakage, or loose terminations.

If there is corrosion, replace and find out why it is corroded (moisture, etc.). If you cannot correct the cause, consider using a humidity and dust proof spring cover jack—see Chapter 8—Surface types.

If a wire is broken, reterminate it.

If a terminal is loose, tighten it.

☐ If you do not get dial tone after these checks, see V.

If Your Trouble Is a Short

(A) Check each jack to see which give dial tone. The jacks

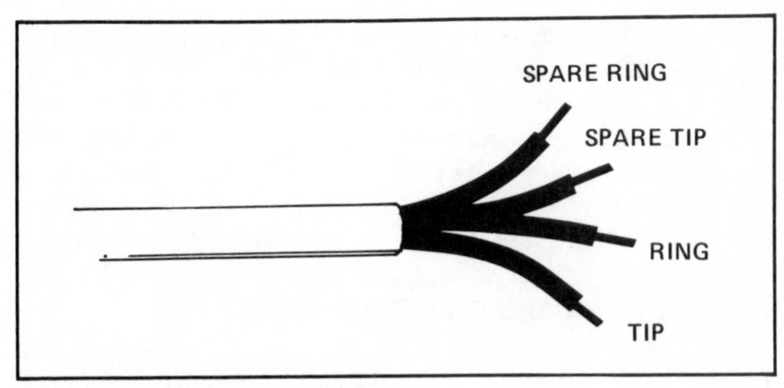

Fig. 11-8. Spare pair in station wire.

connected to the bad wire that you did not reconnect will be dead.

(B) Physically inspect the closest bad jack.

(C) Remove all of the wires terminated to the inside of the jack.

(D) Clip the test set onto each pair of wires. If one pair allows dial tone, that is the pair to the wire junction or protector. This means that the wire is good to this point.

☐ If you get dial tone

Reterminate jack but do not remount jack in wall yet.
Go to next bad jack and repeat procedure.
When the problem is corrected, remount all jacks.

☐ If there is no dial tone

The trouble is on the wire between the wire junction or protector and this point. See V for your options.

V Wiring

There are three solutions to consider: (A) using the spare pair Fig. 11-8, (B) physically repairing the damage Fig. 11-9, and (C) running new wire (see Chapters 5 and 7).

(A) The spare pair (usually yellow and black) need to be made down to the green and red terminals or spadetipped leads. Note: This will not allow any options such as a second line, or power for lights or buzzers, if the wire is four conductor.

(B) Physically inspect the length of wire from the problem jack to the wire junction or protector. Watch for broken or cut or crushed wire, shiny spots, green copper corrosion, or staples cutting into the sheath. Once you have located the problem you can splice the wire (see Chapter 8). This should only be done indoors (outside would only be a temporary solution).

(C) Run a new length of wire from the jack to the wire junction, protector or working jack (closest or most accessible). See Section on Installing Jacks in Homes with Service (Chapter 8).

PHONE GOES DEAD OR DISCONNECTS

I CHECK THE CORD: Wiggle the cord at the jack and at the base of the phone. The cord may be damaged or the wrong kind. If a receiver cord (smaller plug) has been used instead of a mounting cord between the phone and the wall, the phone may go dead intermittently. See Fig. 11-10.

II TEST LINE COMING TO HOUSE: Check this according to the instructions under *no dial tone*. The difficulty in doing this is that since the trouble is intermittent, you must test while the trouble is occurring.

NOISES ON THE LINE

Humming

(Ia) A rare but possible cause of a hum is the close proximity of an appliance. If this seems possible (as in the kitchen) move the phone farther away and see if the trouble persists.

Fig. 11-9. Repair of wire.

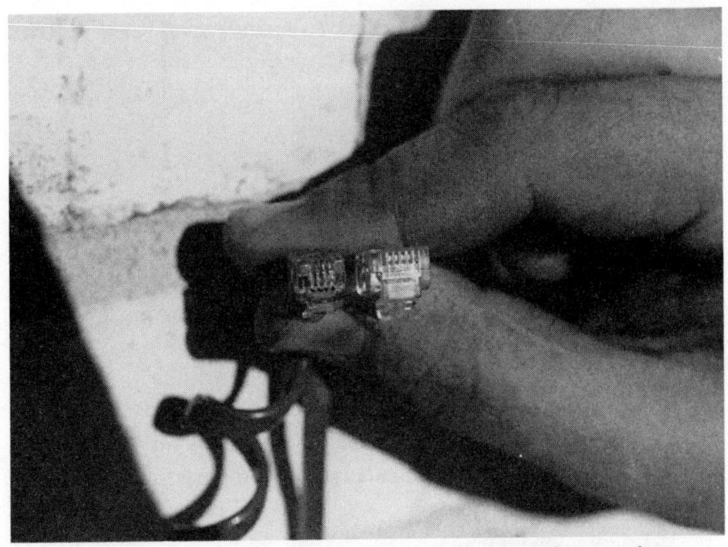

Fig. 11-10. Modular phone plugs, receiver, and base cord comparison.

(Ib) A hum is usually caused by a wire bare of insulation making contact with a ground or ground source.

Determine whether you have an interface or a protector. Check Chapter 2 for information.

Interface

(A) Locate Network Interface
(B) Unplug the modular connection to the interface.
(C) Plug a telephone into the interface.

- ☐ If the hum is still there, the problem is in the phone company's line. Call the telephone company for repairs.
- ☐ If the hum disappears, the problem is in your wiring.
- ☐ Replug modular cord back into interface.

(D) Remove all inside station wires from the wire junction. Connect a testset to the terminals that provide dial tone (red and green). Place each pair of wires on the terminals of the wire junction which is providing dial tone, a pair at a time.

- ☐ The pairs that do not hum are fine.

☐ The pair that still hums is damaged. Go to the *to clear the hum* section.

Protector

(A) Locate the protector.
(B) Remove the cover.
(C) Remove all inside wiring from the protector terminals (use a 3/8-inch socket on handle).
(D) Connect a homemade test set (see Chapter 9) to the protector terminals.

☐ If the hum is still there, the problem is in the phone company's line. Call the telephone company for repair.
☐ If the hum disappears, the problem is in your wiring.

(E) Place each pair of wires to your home on the terminals of the protector. The pair that still hums is damaged. Go to the *to clear the hum* section.

To Clear the Hum

There are three solutions to consider: (A) using the spare pair (B) physically repairing the damage, and (C) running new wire.
See *no dial tone V wiring* for details. Note: There may also be a staple that has been driven into the conductors inside the sheath grounding against wet wood. This hum will show up when it rains.

II HEARING OTHER VOICES WHILE ON THE PHONE

Crosstalk

(A) Crosstalk is caused by the outside lines bleeding over into each other.

☐ If it is only happening when the ground is wet after a heavy rain, you just have to wait and let things dry out.
☐ If it occurs frequently or is very annoying, call the phone company to repair. It is their problem.
☐ The other possibility is that if your home has two separate telephone numbers on the same station wire this may cause crosstalk. Place one of the numbers in a separate station wire to clear the problem.

STATIC

(A) The telephone company line that comes to the house is the most common location for static trouble. You can be fairly certain that this is the problem if it is very bad after a heavy rain and clears up as things dry out.

- ☐ Check interface or protector. See the *no dial tone* section. Test the Line Coming to the House.

(B) Using a receiver cord instead of a mounting or base cord at the wall will cause static because the receiver cord has a smaller plug. Wiggle to see if it is loose. Replace the cord.

(C) Examine the jacks for corroded terminals. If they are corroded, replace the jack.

(D) Examine the telephone for switchook contacts that do not open and close properly. Don't try to repair these yourself. Have a qualified refurbisher repair or replace. See Figs. 11-11A and 11-11B.

(E) Station wire that has been pigtailed (twisted) together can also cause static.

- ☐ The station wire will need to be inspected from the jack to the end where it is terminated.
- ☐ Static of this type will only show up in the one station wire or jack that is causing a problems.
- ☐ Use a jack or connector to reconnect. See Splice in Chapter 8 or Repull wire on Chapter 7.

Radio Signals Over the Phone

(A) When a person receives radio signals over their phone, it is not always the "music on hold" feature of the other party's phone. The phone lines will sometimes pick up the signal of a strong radio station and transmit that signal into your phone. This could also be cross-talk from a line that carries music—see *cross-talk*. Call the phone company to put a filter on the line.

BELLS OR SIGNALING DEVICES

(A) Will not ring

Fig. 11-11A. A set of switchhook contacts shown at center of photo.

☐ There may be too many phones on one line. (Check REN in Chapter 2.

Disconnect some to allow the others to ring.

☐ Ringer or signal is turned down or off.

Fig. 11-11B. Switchhook contacts shown close up.

☐ The bell or signaling device is defective.

Swap part with another phone to determine if this is the problem.

Take it in to: repair, replace part, or replace phone.

☐ Defective switchhook is not making proper contact. Refer back to Fig. 11-11A and 11-11B.

Contacts may be:

(1) Corroded - Look at them. Clean with 400 grit sandpaper.
(2) Jammed - Look at them. See if you can find and remove an obstruction or jiggle them.

Switchook may be jammed. If the spring does not pop up and down you may be able just to replace it.

Jack terminals may be corroded. Do not try to clean them. Replace the jack. Find out the cause of the corrosion. If your climate is damp you might need to use the type jack that has a moisture proof cover (see Chapter 8).

(B) One short ring

☐ Terminals in the jack are corroded. Replace the jack. Determine the cause of the corrosion and correct. If your climate is continually damp, you should use a humidity and dust proof spring cover jack (see Chapter 8—Surface).

Physically examine terminals.
Replace jack.

☐ The phone was left in call-forwarding.
☐ The leads to the ringer or signal are loose in the jack or in the phone network.

Examine the phone network.
Take the phone to a refurbisher or the phone company.

(C) Soft Ring

☐ The bell or signal is turned down low.

Turn it up.

☐ The phone is laying on a soft cushion that absorbs the signal.

Place on hard surface.

☐ The bell or the signal is obstructed.

Physically examine.
Remove obstruction.

☐ Too many extension phones on one line.

Check Chapter 2 for REN.
Unplug some phones.

DIALING

(A) Dial tone stays on

☐ Check the rotary dial to see if the contacts are opening. See Fig. 11-12.

If the contacts are not opening, take the phone in for repairs or replacement.

☐ If your phone has a tone type dial, the wires (tip and ring) may be reversed.

One test of this is if dialing sometimes kills dial tone.
Reverse the tip and ring wires (green and red).

☐ A tone dial phone may not dial out because there is no "tone service" available. It will sound the proper pitches but not break dial tone.

Check with the phone service to see if "tone service" is available. Sometimes, however, the service is on file but not yet available.

Plug in another tone dial phone and see if it works.

☐ The tone dial may be defective. In this case it will break dial tone but not dial out. You may get a fast busy signal when you try to dial out.
☐ If you have universal dialing, the pulses may be set wrong. They should be 10 PPS not 20 PPS.
☐ Sometimes one or more buttons on a universal dial are defective. Try dialing a number that uses different digits.
☐ On a touch tone phone a bad polarity guard can cause you to be unable to dial out. This is rare. Take the phone to a refurbisher.

(B) Dials Wrong Numbers

☐ A rotary dial may have defective contacts inside.

Check to see if they are corroded. Refer to Fig. 11-12.
Clean contacts with 400 grit sandpaper.

☐ A tone dial can be defective.

If it is defective, some buttons will produce a very different tone from the others.
Replace it.

☐ Electronic pulse dialing can be set wrong.

Look to see if it is set at 20 PPS.
Reset it at 10 PPS.

(C) Bell taps or clicks when dialing

☐ This is caused by a defective rotary dial. The contacts in the dial are off or not closing properly. (This is a normal occurrence on some dial phones).

Replace it.

(D) Does not dial out or does not sound DTMF tones (Touch Tone™)

☐ The contacts on a rotary dial may not open and close properly.

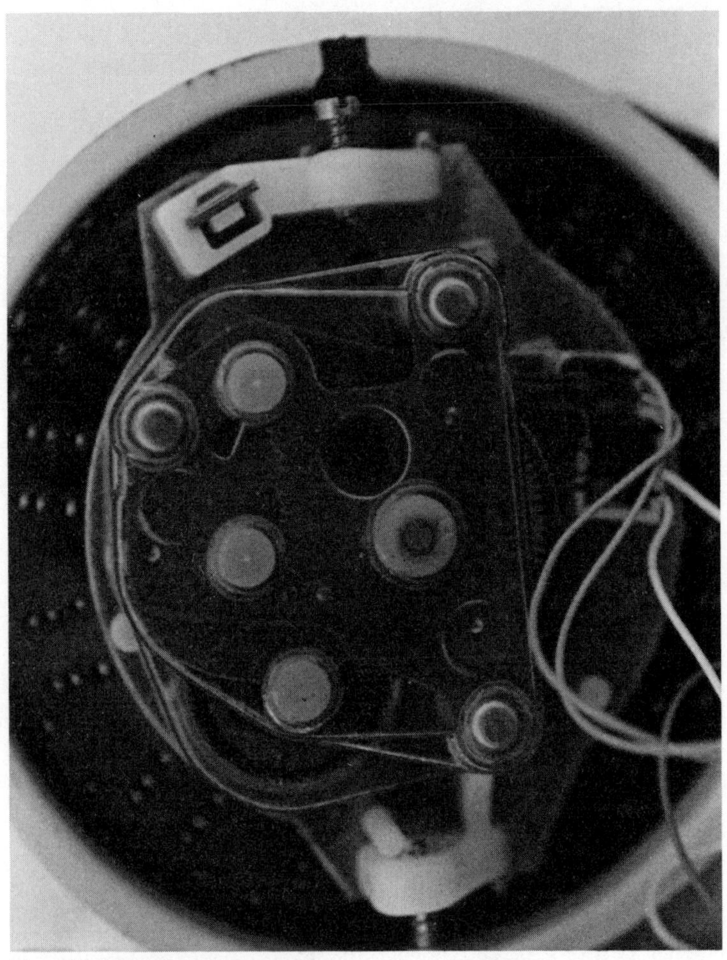

Fig. 11-12. Rotary dial contacts.

Physically examine.
Take to a refurbisher or replace.

☐ Check with the phone company to be certain that you have service for tone dialing. Even if they say you do however, it is possible for the service to be on file but not yet available.

☐ The tip and ring (green and red wires) may need to be reversed.

TRANSMISSION

(Sending and receiving voice signals)

I You Can Hear Caller but Caller Can Not Hear You

(A) A terminal in the phone may be loose. See Fig. 11-13.

☐ Physically inspect.

Fig. 11-13. Inner workings of phone from above.

☐ Tighten loose screws.

(B) The cord on the phone receiver may be defective.

☐ Exchange with a known good cord.
☐ Replace.

(C) The transmitter on the handset may be bad.

☐ Exchange with a known good handset.
☐ Replace.

II You Cannot Hear Caller But Caller Can Hear You

(A) The cord on the phone receiver may be defective.

☐ Exchange with a known good cord.
☐ Replace.

(B) The phone receiver on the handset may be defective.

☐ Exchange with a known good receiver.
☐ Replace.

(C) The contacts in the rotary dial phone may be corroded.

☐ Physically examine.
☐ Clean with 400 grit sand paper. Eliminate the cause of corrosion.

(D) The problem could be a defective switch hook.

☐ Check contacts to see if they are corroded or are not opening and closing properly.
☐ If corroded, clean with 400 grit sand paper. Eliminate the cause of corrosion. If the contacts are not opening or closing properly, you can try to adjust but most likely you will have to replace.

III Other Party Hears Distortion

(A) The line or trunk to which the call was made could be bad.

Fig. 11-14. View of handset phone parts with the transmitter on the left and the receiver on the right.

Hang up and call back for a fresh connection.

(B) There could be a loose connection in the jack or phone network. Refer to Fig. 11-13.

- ☐ Examine the screw terminals inside the phone.
- ☐ If any wires are loose, reterminate them.

(C) It could be a defective transmitter (the part of the mouthpiece with the tiny holes into which you speak). See Fig. 11-14.

- ☐ Exchange with one that is known to work. (It screws off and on.)
- ☐ Replace.

IV Voice of Caller or Person Called Decreases in Strength

(A) The trunk or connections between you and the other party may be bad. Hang up and try again.

(B) Too many phones off of the hook can cause the signal to decrease in strength. This is more of a problem on long distance calls than on local ones.

MAINTENANCE

If your telephone is soiled you may wish to clean it.

(1) You may clean a phone with a glass cleaner. Be very careful not to get the solution inside the inner components. If you think that the solution might harm your phone, place a small amount in a hidden area before using it on the rest of the phone.

(2) If the phone is badly soiled or scratched there are telephone cleaning products that can be used to help refurbish the shell.

(3) If you peel paper stickers from a phone but the adhesive does not come off:

(a) Place a small amount of either LPS® or WD 40® on the area. Do not get the lubricant on the inside components of the phone.

Wait a moment or two.

Rub the adhesive from the surface.

Spray glass cleaner on the area to take off the lubricant.

(3) To remove the face plate or fingerwheel on a dial phone or rotary type:

(a) Unbend a small paperclip.

(b) Turn the dial all the way around as though you were dialing the operator.

(c) It will stop after it has passed the finger stop.

(d) Push the end of the paperclip into the small hole at the bottom right of the dial to force a small lock tab down while you are applying slight pressure by turning clock-wise. See Fig. 11-15. The fingerwheel should release. If not, repeat.

(e) Slide the fingerwheel from under the finger stop and remove it from the face of the dial.

(f) Clean the dial and the fingerwheel.

(g) Redesignate the card and place it back on.

(h) There are a couple of protrusions at the edges of the fingerwheel to help in mounting the card. Take advantage of them.

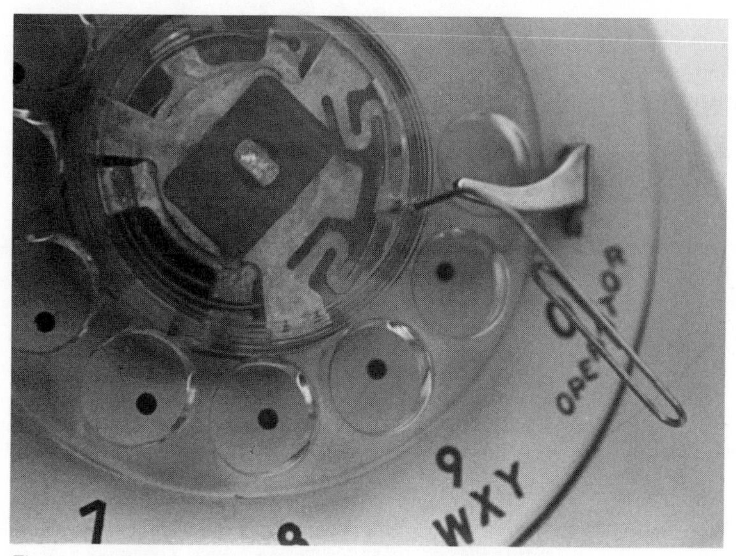

Fig. 11-15. Fingerwheel with unbent paper clip inserted.

(i) Place the finger hole which would be 0 digit over the 9 digit.

(j) Turn the fingerwheel counterclock-wise as shown in Fig. 11-16. The fingerwheel should snap into place.

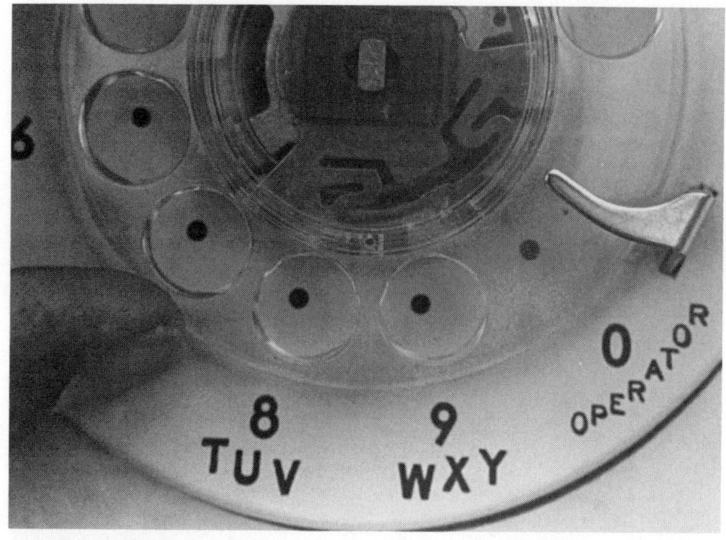

Fig. 11-16. Fingerwheel being replaced on face plate of phone.

Fig. 11-17. Protector showing spider webs that accumulated on terminals.

(4) Keep the phone cords loose, without tension at the ends, and untangled. Pressure can cause the connections to give trouble.

(5) When using the phone, do not pull or strain the cords to reach an area. Install a longer cord or install an extension jack where needed.

Fig. 11-18. Protector being cleaned of spider webs with bristle brush.

(6) Do not let the phone come in contact with water or be in a damp or humid environment for any extended period.

(7) Do not try to disassemble your phone for modifications.

(a) There is very little modification you can make on a (unless it is for business applications).

(b) You may void your warranty if you own the phone.

(c) If the phone is leased, the company may not repair or replace it if you have tampered with it.

OUTDOOR EQUIPMENT

Spiders love to build webs inside of protectors. These webs hold moisture and eventually the termination screws corrode (see Fig. 11-17). Remove the webs periodically with a 1/2-inch bristle brush like the one used to paint trim (see Fig. 11-18).

Chapter 12

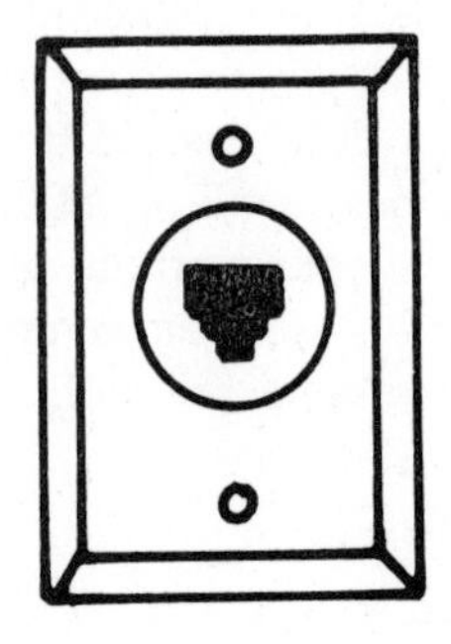

Telephones—Types and Selection

In addition to the right to install your own telephone wiring, the changes within the phone company have given you the opportunity to purchase your own phone. Although you may still rent phones, many people choose to purchase them. You may decide to buy yours in order to save money in the long run or to use special options not available in rental. This section will help you to consider your particular needs and to choose the telephone to suit them. The basic types of phones will be discussed as well as the special features available with them. I will categorize phones as (1) Disposable, (2) Cordless, (3) Standard and Decorator, (4) Multipurpose, and (5) Two-Line Pick up. I will note briefly the special feature types and the optional equipment that can be added to an existing phone system. The final section will include considerations in the final purchase, such as warranty and availability of parts.

DISPOSABLE TELEPHONES

"Disposable" is a nickname for the least expensive phones you can purchase. They are also called "chirper" or "tweeter" phones because the transmitter or earpiece also serves as the ringer. The resultant ring gives them this name. A "disposable" may run from twenty dollars down, depending on the available features. The phone will usually pay for itself within a year. If problems occur, it is often simpler to dispose of it than to attempt to have it repaired.

The difference between this new electronic phone and the old Bell type phone (now AT&T) is in the design. The older type phones were made with a network with several loose wires connecting on screw terminals. The new phones are made with printed circuit boards. The dials are like the key pads of a calculator. The carbon microphone of the older phone has been replaced by electret ones. Cords on the disposables are permanent and cannot be replaced.

Some of the features that were not available with the old-style phones are common with "disposables." These include: last number redial (a convenient way to continue dialing a busy number while working at another task), memory dialing (which dials selected frequently dialed numbers at a touch), mute button (prevents transmission of sound to the other party without cutting them off), ringer volume control (allows you to adjust the ring to be heard at some distance or to be least disruptive). This makes the phones a lot more attractive than the old style phone but you may need to check some items on a selection of this type of phone.

The placement of the switchhook is important since when using one of these phones can cut you off in the middle of a conversation if you accidentally grab the instrument where the switchhook is located. Then also the switchhook may not always hang up properly when it is set down on a flat surface and this has to be on a flat surface. The best way to hang this phone up is to use the wall holster that is provided.

There are also times where you may need to place the phone down for a moment. What usually occurs is the phone gets pulled to the floor by the cord; the wall holster allows the phone to face toward you while holding it out of the way. If the phone does not come with one, you may want to consider it again. A problem with the wall cord on these phones is common as with all modular plugs is the tendency for the squeeze tab on the plug to break off if the phone is moved around from jack to jack. Once the tab is broken off the phone can not be plugged in properly. This leads to disconnections while on the phone or the phone not functioning when it is assumed it is connected in the outlet. On a cord that is permanent this can become a major problem since it can not be easily replaced with another. At the end of the section I will explain options for this problem.

Some phones of this type may not have the dial in between the transmitter (mouthpiece) and receiver (earpiece) but on the back of the phone at the top. This will not allow the phone to be placed between the cheek and shoulder without cutting you off from the conversation. For some people who may want to be on the phone for a longer period of time and have their hands free this will not do. The length of the phone is of importance also when conversation is not clear. The distance between the earpiece and mouthpiece should be at least 7 1/2 inches. This will allow you to speak directly into the mouthpiece while the earpiece is against your ear. One further note is that these phones have a sort of peculiarity. When another phone is dialed with a pulse dial the dialing is heard on this type of phone. The disposable phone will make a fine extension phone for the home but I would not favor it as the main phone of the home.

CORDLESS TELEPHONES

The need for mobility of a phone in different areas of the home and also outside has made the cordless phone a desired item. It is not just placed for this purpose but also the use for people that are ill and not able to move around the home very well. The cordless allows the use of a phone without the problem of only having certain places to plug into. The cordless is made of two separate units, the phone unit or handset and the base unit or station. The base unit is the portion that is connected to the actual telephone jack and the ac power. This unit makes it possible to operate the phone without the need of the external connections.

The basic principle of a cordless phone is the use of FM radio waves between the base unit and the phone. The phone is powered by a set of rechargeable Ni-Cad batteries that are recharged in the base unit every time it is necessary. The recharging depends on the style of the phone and the amount of operation. The versatility of this phone is its mobility and the distance it can be taken from the base unit.

Though this seems like a perfect phone to some there are some problems that need to be gone over. The phone itself may operate in the same way a typical phone does but all its signalling is controlled by radio waves. These are interfered with by other radio waves and sometimes the base unit could be pirated by other people with the same phone unit. This has been corrected somewhat by the use of code numbers on the phone and base unit to allow no one to gain easy access to your phone line. This coding system seems to be working well. The phone unit will not always give the same reception as the typical phone connected to the jack in the home. The problems with the use of the cordless is also multiplied by any large metal structure in the way or the use of metallic wall coverings as wall paper. This can limit or completely distort the signal. The use in a metal trailer is not a good idea. In areas of large population as in an apartment complex the problem of other phones being on your frequency is greatly added to the possibilities of interference. The area may have high electrical induction and electrical interference to cause problems. The FM signal is used to prevent this type of problem but it can still come up.

A physical problem I have noticed on certain cordless phones is the Talk/Standby button on the handset unit. This button is what places the phone in position to receive a call (Standby) or make an outgoing call by receiving the dial tone (Talk). The buttons on some of the phones are in very uncomfortable positions or not well suited to switch from one position to another. Check this when purchasing a cordless. The one item that you should remember when you actually own a cordless is the signal the phone gives when receiving an incoming call. This signal will keep going on as long as the phone is on Standby mode. Be sure to switch to the Talk mode before placing the phone to your ear. The signal may cause hearing damage. Cordless phones come with other features much like the disposables such as redial, mute, memory, and other forms of electronic advantages. The ways to select a cordless will be discussed at the end of this section.

STANDARD AND DECORATOR SETS

The standard phone that has and still is being used by many households is the same type that was and is leased to people now by AT&T. These phones have no special calling features on them. They can be either DTMF or pulse dialing on the set. Many of these are still hardwired to the existing jack and have not had any problems for many years. The phones were constructed with the intent of not failing in service. These phones can be converted over to a modular cord if the old cord is the kind with the spadetips on the end. This is done by removing the old cord from the set first by removing the outside shell or cover via two screws at the base of the phone. Next remove the existing cord from the screw terminals on the telephone network. These will be the L1, GN, and L2 terminations on the network. If simpler for you, match the colors of the new cord color by color; this would save the problem of not finding the correct terminations. The new cord should be one with spadetips on one end and a modular plug on the other. When converting the cord be sure to hook the metal clip that holds the cord to the base. This clip keeps the cord from being pulled from the terminations and pulled from the clip. Remember this modification should be done only on a phone you own, not one that is being leased.

The standard phone is able to have calling features even if they are not on the phone. The new ESS (Electronic Switching System) allows a person to do some pretty amazing things such as *call waiting* where you are able to put one party on hold while you pick up the other one. There is a tone on the phone while you're speaking to the other party to signal that someone is trying to get in touch with you. With *call forwarding* you are able to transfer your telephone number to ring at another telephone number as if you were home. *Speed calling* makes it possible to have a code number for certain frequently called numbers without having to dial the whole number. *Three way calling* allows you to talk to two separate parties at the same time, this allows a second dial tone to call the second party. This is all done on a standard telephone or any out dialing telephone that has an area supplied with ESS dial tone.

The modernized types of standard phones are decorator models. These phones are more in line of being conversation items but are true operating phones and can be very serviceable. They come in many shapes and sizes, from Coke bottles, red painted lips to very fine woodgrain cabinets. The price range of these phones can be modest to highly expensive depending on style. Decorator phones are functional but they are not always built with the durability of the standard no-frills phone. The phones have certain troubles as with the construction of some of the parts. This is mainly due to the fact that some of the decorator phones are built around some object that was not intended to be a phone. Also when some phones are duplicated by the use of plastics they can not always take the same abuse as the original model. Two items that come to mind are the areas where the handset or receiver hang up on the switchhook. There is a tendency for this part of the phone to malfunction or even completely break. The cradle (the part of the phone which holds the handset) on certain phones will also suffer the same fate. When buying any decorator check for these items for durability and also other places where there may be possibility of breakage from the constant use. The decorator in my opinion should be a secondary phone to a good main telephone.

MULTIPURPOSE TELEPHONES

The ability to purchase phones in any style approved by the FCC has encouraged the manufacturing of phones with other functions. These other functions included with the phone itself are: clock-radio, answering machines, cordless phones with clock-radio, intercoms, and speaker phones. Some items such as the answering machine can be considered a portion of the phone itself. All these features add to the possibility of being able to have everything in one compact unit in one area. When going over these features check to see what would be the most practical for you to use when you do finally make the purchase of a multipurpose phone. When this is decided check to see if the phone will require the need of ac power to operate any of the features the phone has, this is usually always the case. Be sure the area you decide to place the phone will be in easy access of a phone jack and ac power.

It is difficult to say what type of problems to look for in a phone of this type. My best advice is to check items that are true of any good phone such as the sturdiness of the case and whether the cord is easily replaceable. Then as always, check the warranty and what policy the store has about items you are not satisfied with as far as returns.

TWO-LINE PICK-UP TELEPHONES

Telephone use in the home has increased drastically in recent years. At first people were able to accept party lines then more people were getting private lines. This trend has come to where consumers are wanting two private lines working in a residence instead of the one line. This additional line is for the children or for a business in the home, or simply the need for an additional number. This changing trend in the home has brought about a more desired phone that eliminates the need to have two phones in one area. The clutter of two phones plus the problem of having to tell which line is ringing or worse getting up to pick up a different phone in another part of the home makes the desire for one phone to pick up two lines a much wanted and needed item.

This type of phone is also able to work on just one single phone jack because it is able to carry two separate telephone numbers. The USOC code on this jack is an RJ14C. This is the same jack you would plug any standard one line phone into. The difference is that the spare pair of contacts in the jack will carry the additional line for your two-line pick-up phone. The two inner contacts would be for the first line while the two outer contacts would be for the secondary line. Since most wire now used in the home has four conductors this makes it possible to have this arrangement. It is also possible to pick up both lines from different jacks into the phone by the use of an adapter that plugs into the phone and plugging both modular cords into the adapter.

The two-line pick-up phone is also handy and functional. The features and style can run from a trim little decorator model to a style much like the electronic business phone in an office. Because of electronics and the two lines, it can be a fairly reasonable business phone without the need for a control unit or KSU (Key Service Unit). This is a unit needed to operate business phones and helps in routing calls. On the two-line pick-up you will probably not need over two or three phones of this type to operate the two lines effectively. If you need one not on a table or desk there are wall models you may hang on a wall mount jack. Certain desk models can be converted into a wall mount phone by changing the back plate of the phone.

With these phones you can get features such as, hold on one line while calling the other, conference calls where you can call or talk to two parties (3-way conversation), distinctive ringing where you can tell which line is ringing, LED lamps to tell if a number is in use or on hold, speaker phone, memory dialing or redialing when the party's line you were calling is busy, pause button, mute button, and pulse or tone dialing. This shows that these phones are meant to offer the consumer some more versatility with the two lines. One more point is that you can have the numbers rotate from one main number to the second line when the main is being used. The only problem is to be certain that the first three digits in the numbers match. This will enable you to rotate from the main to the second line, not the other way around.

If you feel that the purchase of these phones is a bit out of your price range, and you feel you still wish to have this option, there are adapters that allow you to plug a standard phone into them. To pick up two lines you plug the adapter into the jack or jacks. Then plug a standard phone into the adapter. These adapters can make a link for the two lines in the home without having to buy a whole new phone.

When looking for these types of phones your best bet would be in telephone specialty stores that carry a wider range of phone items such as the two-line pick-up phone. One final note, there are cords that only supply the first line to the phone but not the second line. The modular cord will only have two contacts in it. This type of phone requires four contacts.

TELEPHONE SELECTION

Now that it has become a must to have your own telephone and to maintain it, you have to find what is your best buy and the best choice for you. This, like the buying of any good appliance, requires some planning and effort to decide if the product you purchase will provide you the most satisfaction. I hope to show some insights and pitfalls in selecting a phone that will give you good service as well as providing the features you wish to have. These are what in my judgement allows you to decide if a phone is desirable despite some items that may or may not cause problems later.

To begin with, when buying a phone or any other item, there are certain tips you need to remember. First is to check what the store policy is on returning an item. Can it be returned after a reasonable time? Do they allow money back or credit on another purchase? Does the company have a repair department or ship the item back to the manufacturer for repair if the item can not be exchanged? If it is shipped back do they let you have a loaner while your item is being repaired? Check the warranty for how long it is and what it covers as far as parts and labor. Is the company you are buying from one that you can reasonably say will not fold or be unable to honor its warranty. Though this may seem like a lot of questions they have a bearing on whether the item you buy will be a good purchase or just another headache. Another point to check is the price of the item at different stores. It may be cheaper in one place, but they may not service the item at all. If this is the case see if they will exchange the item if it is faulty or defective. In the long run the price may be less, but the product and service may not come up to your expectations.

Here are some things to check when selecting a phone you wish to purchase. Some of these can be inspected at the store, others such as the quality of voice and signaling need to be checked at home. You should test your phone at the store before you have to take it home. In helping you to do this I have separated the selection of phones into two parts. First is a general inspection at the store, and second, inspection of the phone's operation at home, if you are not able to check it at the store.

GENERAL INSPECTION

(1) Does the phone allow the cords on the handset or the base to be replaced easily if damaged later on? This is most important for the base cord. The base should have a modular plug in for a new or longer cord. For a permanent cord to a phone base, a good thing to check is the plastic sleeve that holds the base cord to the set. This helps take tension off the cord at the phone connection, which may lead to problems.

(2) Do the cords plug in firmly but not so tightly that it is difficult to disconnect? The modular plug should not have any play sideways.

(3) When the phone is shaked does it make sounds of some parts being loose or not cushioned? This may mean trouble before you start.

(4) Do the cords seem short and not allow much movement for the user? The base and handset cords should be a minimum of 5 1/2 feet long.

(5) Is the phone easily portable if meant to be? This excludes the multipurpose and cordless base unit which are meant to be stationary because of being plugged into ac.

(6) Does the phone look flimsy? Does it have any attachments that may allow easy breakage? (Example: the cradle of the phone that holds the handset is weak and does not look like it can survive much use.) Is it attached to the rest of the phone, by a simple rod or by molded plastic? Slamming a phone down hard or carrying it around will cause some phones to break in this area.

(7) Is the base of the phone so light that when the handset cord is pulled by some of its tension in the coils it will pull the base off the table? This is usually true of phones where the main components are located in the handset and not the base.

(8) The distance between the mouthpiece and the earpiece (approximately 7 1/2 inches) is important for conversations to be heard properly.

(9) The receiver or handset should depress the switchhook easily. In some rare cases the handset is not of sufficient weight to operate the spring action of the switchhook.

(10) The dial should feel comfortable for the people who are going to use it. Some consumers with large hands find the smaller key pads and dial type phones difficult to operate.

(11) Will the phone be wall mounted or a desk model? Some

desk models can be converted to a wall mounted instrument.

(12) Is it comfortable to place the phone handset between the cheek and shoulder to allow free use of the hands? Some handsets do not allow this because they are shaped oddly. Some have a dial in the back of the handset. When it is held between cheek and shoulder it cuts off the conversation.

(13) For certain hard of hearing consumers, hearing aid compatibility is an important consideration.

(14) Check to see if the phone you are going to purchase is legally registered with the FCC. Note: FCC telephone registration does not have anything to do with the quality of a telephone instrument or equipment. It only indicates that the equipment will not damage telephone lines.

(15) Will the phone be compatible with the new long distance services now being provided? A misconception about the new key pad dials is that they are DTMF or Touch Tone®. Although they are push button, they do not always give tones. Check for this before the purchase.

(16) Does the phone you are going to purchase also require ac power to operate? If so, do you have a phone jack close by in the area from which you wish it to work?

(17) Some phones require a battery to operate certain features, such as memory. If so, can the battery be replaced easily by you or is it a special type that may be hard to obtain?

CHECKING TELEPHONE FUNCTIONS AT HOME

(1) When you lift the receiver from the hook, does it give a good clear dial tone?

(2) Call someone. Is the volume loud enough for conversations to be heard clearly? Do they hear you as clearly as you hear them?

(3) When you call someone is their voice recognizable? Is the sound quality acceptable for both you and the other party?

(4) Check the sidetone in the phone. The sidetone is your ability to hear your own voice back on the receiver so you will not feel you have to shout into the handset to be heard.

NOTE: When you are testing these qualities (1, 2, 3, 4,) on a phone remember that if other phones are off the hook at the same time there will be a large drop in the level and quality of sound from the phone.

(5) Have someone call you back. Listen to the ringer or signaling of the phone. Is it too soft or too loud for your needs? You can adjust some phone ringers to be either loud, soft, or off. Try the adjustments.

(6) There may be problems with interference when you operate a cordless phone. This can include hearing other people on your own frequency while you are using it.

(7) Speaker phone setups on phones can give bad voice reproduction. A common fault is that people sound like they are speaking in a well. Check the quality from the receiving party you have called. Also remember this feature is to be used close and not across the room, which will make the sound quality drop greatly.

(8) The last major task is to see that all the features of the set you have bought function correctly and consistently. These features may be one main reason you purchased this particular phone.

TIPS FOR MAINTENANCE OF TELEPHONES

A problem with the modular plug is that the squeeze tab may break when you are moving the plug from one jack to another. To help prevent this on permanent cords (wall or base) you can place an in-line coupler on the plug, then a separate base cord to plug into the jack. This cord does not have to be long since you can get short one foot cords to do the job. Prevention is better than replacement or repair.

If the squeeze tab on the cord is already broken and it is a permanent cord you can replace the modular plug with a modular crimping tool. This tool is not expensive (around $10.00 not including the modular connectors). The original intent of this tool is to make custom modular cords, but works well to repair broken or damaged plugs. Before you work on the cord or any portion of the phone remember that anything that you do to a phone will void the warranty.

The modular crimping tool has directions for its use. Follow the tool instructions carefully. On the plug part going into the jack, the black wire should be on the right while the yellow is on the left (when you look at it from the top where the squeeze tab is).

For some more maintenance tips on phones check Chapter 11. If you desire more information on phones that are in style and more of the accessories available to you as a consumer check the TAB books *All About Telephones*, 2nd Edition by Van Waterford (TAB book number 1537) and *The Master Handbook of Telephones* by Robert J. Traister (TAB book number 1316). These books give more detailed information on phones and the features that are available to the public.

Glossary

adapter—A plug-in device to convert one type of connection to another type. On phones this can change a four prong to modular plug or a singular modular plug to a double connection plug.

bonding—The condition where a mobile home is grounded with all electrical systems.

cable—(In this guide 25 pair or more in a sheath). Inside it is called inside wiring cable. Outside it is called "cable."

circuit—The path of electrical flow from the power source through the jack to the ground.

color code—The color identification of each conductive wire to trace from one end to the other.

conduit—The rigid or flexible tubing through which wires are run.

conductor—The wire or anything else capable of carrying an electrical current.

connecting block—The termination device (usually screw lugs) where the conductors join the plug-in or hardware receptacle.

contact rails (also known as modular connector rails)—These are the small wires in a modular jack and the outside of a modular plug. They form for electrical connection when joined by plugging.

continuity—When a circuit is able to carry a continuous electrical current.

conversion—The changing of a receptacle from one type of jack to another; usually hardwire to modular or four prong to modular.

cross—The condition that occurs when the tip or ring of a circuit contacts the conductors of another circuit causing phone problems (static or hearing other phone lines).

crosstalk—The condition that occurs when you are able to hear other conversations on the line that are not the party you are speaking to.

demarcation point—See **point of demarcation**.

dial tone—The signal given on phones to show that the phone line is in proper order.

disposable—The least expensive telephone, which is better to discard than to repair.

drip loop—A downward dip or bend in station wire to draw moisture from an entrance hole.

dropping a wall—The act of concealing wire inside the wall so it can not be seen.

drywall (also known as sheetrock)—A molded preformed wall board usually made of gypsum.

DTMF (*D*ual *T*one *M*ulti*f*requency)—A method of dialing a telephone number by tones.

ESS—The abbreviated term for *E*lectronic *S*witching *S*ystem (Bell System).

extension—Any phone or equipment following the main equipment.

FCC registration number—A number on all phone and phone equipment usually fourteen letters and digits long. This number is accompanied with the ringer equivalent number (REN). These numbers are assigned by the FCC to show equipment is approved for use on telephone lines.

fire wall—The horizontal studs in the wall to prevent fire from spreading rapidly.

fishing—The process of getting wire through finished walls or ceilings.

fish tape—A long spring steel strip used to fish wires or pull wire through conduit.

four prong (or four pin)—A connector with four prongs that plug and unplug easily. This is one of the portable connections for phones.

flush jack—A jack that lies flat against the wall rather than protruding.

ground—A conductor of an electrical current connected to the earth at some point.

hardwire—The connecting of wire permanently to a receptacle with spade tips or bare wire where it can not be easily disconnected.

hex-head screw—The type of screw that has a six-sided head.

installation—Any service that allows the use or future use of phone equipment.

insulation—The nonconductive covering that protects the wires.

jack—A telephone receptacle.

joist—The cross beams that run wall to wall.

line—One phone number.

made down—Same as terminate.

modular—The small universal connector that is used to plug or unplug a phone or phone equipment quickly.

Molly® bolt—The more common term for a B wall screw anchor or hollow wall anchor.

mounting cord—The connecting wires from the wall receptacle to the telephone base (sometimes referred as a line cord).

network interface—The equipment of demarcation where the phone company's equipment ends. This equipment is where a person plugs in to get the dial tone (also referred as interface).

noise—Any outside interference such as a humming or buzzing sound.

non-flush jack—Jack that protrudes from the wall.

open—The break in the wire where the flow of electrical current stops.

outlet box—Usually a prewiring box that can have a jack installed in it.

pair—Two conductive wires that make an electrical circuit.

panhead screw—A screw with a circular head (like a pan bottom).

party line—A phone service that allows more than one party to use a pair of wires.

pigtail—Joining two or more wires by twisting together to form a splice.

polarity—The description of voltage being negative or positive from a given point.

point of demarcation—The location where telephone company equipment ends and the customer's begins.

prewire—See **rough-in**.

protector—A type of telephone fuse box used to divert hazardous power from inside phone equipment. It may be installed inside or outside. An outside type is enclosed in a plastic or metal housing.

pulse dialing—A method of dialing a telephone number by interrupting the phone current (as in a dial phone).

push button—A phone that can dial out by means of a numbered button being pushed. Note: there are phones that can dial out on either tone or pulse service because they convert the numbers to pulse dialing. You can dial out on a tone service with pulse dialing but not the other way around.

quad—Term given to any telephone station wire that has four insulated conductive wires in a sheath.

receiver—The portion of the telephone handset on which you hear the party talking. It has been referred as the handset.

receiver cord—The connecting wire from the base of a phone to the handset or receiver.

refurbisher—A FCC approved company that repairs or reconditions telephones.

resistance—Hinderance of the flow of an electrical current. This is the base unit of Ohm's law.

REN—*R*inger *E*quivalence *N*umber.

ring—Battery side of the phone circuit.

ringing current—The low amp ac power surges sent on the phone line to ring the bells or other signaling devices on phone equipment.

rotary—A dial phone.

rough-in—The wiring of a building while it is under construction. Another term is prewiring.

run (or running)—The act of routing wire.

Scotchlok® connector—A solderless connector that crimps two or three wires to make a splice.
shiner—A wire conductor with some amount of copper showing through the insulation.
short—A condition where the tip and ring of the circuit make contact with each other.
sole plate—The bottom horizontal portion of a wall usually a 2-by-4.
spade tips—The flat connecting ends of a telephone cord or jack that slip under the heads of a screw lug.
spare pair—The conductors of telephone station wire that are not used but held for future use.
splice—The bridging of wire conductors.
split-beam—A termination device that cuts through the insulation of a conductor to make contact, thus eliminating the need to strip the insulation.
squeeze tab—The tab on a modular plug that secures it inside the jack.
static—The crackling or popping noise on a telephone line.
station wire—The sheathed wires used in wiring telephone jacks. One common type is quad (four conductor).
stripping—Removing the insulation or sheath from the wire.
stud—The vertical beam in a wall.
switchhook—The spring device in a telephone that closes switch contacts to complete the telephone circuit for sending and receiving. When open it will receive a signal.

Teflon—Stiff sheathed telephone wire that does not emit fumes when it burns.
terminals—Metal connectors on connecting blocks or protectors for terminating wires.
termination—The joining together of the conductors of a circuit on a terminal.
testset—A device to check for the dial tone or problems on a phone line.
tip—The ground side of a phone circuit.
top plate—The top horizontal beam of a wall, usually a 2-by-4.
Touch Tone®—AT&T trade name for DTMF.
transformer—A device that reduces or increased voltages.
transmitter—The portion of the telephone handset that converts speech to electrical signals. This is what you talk into.

USOC—*U*niform *S*ervice *O*rder *C*ode.

wire—The conductor through which the electrical current flows.

wire junctions—A termination device for adding extensions or connecting inside station wire to an interface.

Index

Index

Edited by Roland S. Phelps

Other Bestsellers From TAB

☐ **66 FAMILY HANDYMAN® WOOD PROJECTS**

Here are 66 practical, imaginative, and decorative projects . . . literally something for every home and every woodworking skill level from novice to advanced cabinetmaker, room dividers, a free-standing corner bench, china/book cabinet, coffee table, desk and storage units, a built-in sewing center, even your own Shaker furniture reproduction! 210 pp., 306 illus. 7″ × 10″.

Paper $14.95 Book No. 2632
Hard $21.95

☐ **DESIGNING, BUILDING AND TESTING YOUR OWN SPEAKER SYSTEM . . . WITH PROJECTS —2nd Edition—Weems**

You can have a stereo or hi-fi speaker system that rivals the most expensive units on today's market *. . . at a fraction of the ready-made cost!* Everything you need to get started is right here in this completely revised sourcebook that includes everything you need to know about designing, building, and testing every aspect of a first-class speaker system. 192 pp., 152 illus.

Paper $10.95 Book No. 1964

☐ **INTRODUCING CELLULAR COMMUNICATIONS: THE NEW MOBILE TELEPHONE SYSTEM —Prentiss**

Not just a future concept, cellular communications is already beginning to be put into service around the country. And now, this exceptionally thorough new sourcebook lets you in on what this system is, how it works, and what it means to the future of telephone contact. It contains detailed descriptions of equipment and systems! 224 pp., 90 illus.

Paper $10.95 Book No. 1682
Hard $14.95

☐ **HOW TO TROUBLESHOOT AND REPAIR ANY SMALL GAS ENGINE—Dempsey**

Here's a time-, money-, and aggravation-saving sourcebook that covers the full range of two- and four-cycle gas engines from just about every major American manufacturer—from Briggs & Stratton, to West Bend, and others! With the expert advice and step-by-step instructions provided by master mechanic Dempsey, you'll be amazed at how easily you can solve almost any engine problem. 272 pp., 228 illus.

Paper $10.95 Book No. 1967
Hard $18.95

☐ **BUILD A PERSONAL EARTH STATION FOR WORLDWIDE SATELLITE TV RECEPTION— 2nd Edition—Traister**

You'll find a thorough explanation of how satellite TV operates . . . take a look at the latest innovations in equipment for individual home reception (including manufacturer and supplier information) . . . get detailed step-by-step instructions on selecting, assembling, and installing your earth station . . . even find a current listing of operating TV satellites (including their orbital positions and programming options). You'll also discover that the costs of satellite receiver components or kits are far less than you've imagined! Even shows how to save by using surplus components. 384 pp., 250 illus.

Paper $14.95 Book No. 1909
Hard $21.95